# Total Quality Management

## The key to business improvement

Second edition

A Peratec executive briefing

## CHAPMAN & HALL

London · Glasgow · Weinheim · New York · Tokyo · Melbourne · Madras

**Published by Chapman & Hall, 2-6 Boundary Row, London SE1 8HN**

Chapman & Hall, 2-6 Boundary Row, London SE1 8HN, UK

Blackie Academic & Professional, Wester Cleddens Road, Bishopbriggs, Glasgow G64 2NZ, UK

Chapman & Hall GmbH, Pappelallee 3, 69469 Weinheim, Germany

Chapman & Hall USA, One Penn Plaza, 41st Floor, New York, NY10119, USA

Chapman & Hall Japan, ITP - Japan, Kyowa Building, 3F, 2-2-1 Hirakawacho, Chiyoda-ku, Tokyo 102, Japan

Chapman & Hall Australia, Thomas Nelson Australia, 102 Dodds Street, South Melbourne, Victoria 3205, Australia

Chapman & Hall India, R. Seshadri, 32 Second Main Road, CIT East, Madras 600 035, India

First edition 1991
Reprinted 1991, 1992
Second edition 1994
Reprinted 1995

© 1991 Production engin...

© 1994 Peratec Ltd

Typeset in 10/12 Palatino...
Printed in Great Britain by...
Suffolk

ISBN 0 412 58640 1

A Catalogue record for this book is available from the British Library

∞ Printed on acid-free paper, manufactured in accordance with ANSI/NISO Z39.48-1992 and ANSI/NISO Z39.48-1984 (Permanence of Paper)

# Contents

# Preface to the first edition

Quality is a customer issue. It arises because customers require products and services, which not only meet their performance requirements but are satisfactory in terms of safety, length of working life and pride of ownership. In a manufacturing organization, therefore, the achievement of quality standards is not restricted to the production departments. It extends to all parts of the business from conceptual design to marketing, from order processing and distribution. A quality product is not just a solidly made item dating from the days when 'Made in Britain' distinguished goods from all the inferior products coming out of the emerging industries of the Far East. It is a product which ranks high against all the criteria which sophisticated consumers now use to evaluate the things they buy.

If you agree with the argument that a company is much more likely to produce high quality if all departments are motivated to achieve high quality results then you already have a good understanding of the basic principles of Total Quality Management (TQM). But TQM is not a 'quick fix' or a magic cure. It is a management technique designed to involve all parts of the business in the pursuit of, and commitment to, the highest quality result.

By involving everyone from the Chief Executive to the most junior employee in the company's objectives, in a way which means something in their particular job, the company is well on the way to achieving the best results its workforce can achieve.

Quality standards imply, of course, an understanding of what the customer really needs. In TQM every person in the company should recognize that they have customers even though they may be internal customers. This concept of customer satisfaction provides a basis for establishing competitive measures, performance targets, better communications and, in consequence, a process of continuous improvement. This is the real aim of TQM and the only basis on which companies will achieve world-class performance in the 90s.

I hope this book gives you food for thought. You may not agree with all the ideas expressed but, if it encourages you to consider how Total Quality might be implemented in your organization, it will have been well worthwhile.

Ronald A. Armstrong
Chairman, Peratec Ltd

# Preface to the second edition

It is clear from the success of the first edition of this book, launched in 1991, that Total Quality Management (TQM) is widely accepted as the fundamental business issue of the 1990s and the key to business improvement.

The many positive comments made about the book have been most gratifying. However, as in all things, we have found room for updating and improvements.

In particular, we have revised the sections on Hoshin Planning and Benchmarking to reflect suggestions received and requests for more detail on these important subjects. There has also been considerable updating of detail and new developments – like the European Foundation for Quality Management (EFQM) European Quality Company Awards – have been included.

We have added three new case studies to Part Four – Profiles of Success Through Total Quality – to further illustrate the benefits of the Total Quality approach. The remaining profiles have not been updated because the experiences of the organizations are still relevant and the lessons to be learned are still valid. However, the reader should remember that, since 1991, all those organizations will have moved on, in accordance with their continuous improvement philosophy.

This second edition incorporates all of this in a new paperback format to make the book more accessible, more usable and, we hope, even more popular than the first edition has been.

Paul Spenley
Director
Peratec Limited

# Part One

# Introduction and Executive Summary

## Introduction

This executive briefing report considers the strategic aspects of what is the key management issue of the future – Total Quality Management.

So . . . what is Total Quality Management?

It is a philosophy of management that strives to make the best use of all available resources and opportunities by constant improvement. Total Quality Management is the key business improvement strategy and the key management issue of the future because it is essential for efficiency and competitiveness. The case for Total Quality is made in a later section, 'Total Quality . . . is it really a strategic issue?'

Part Two describes the concepts of Total Quality and provides guidance on how they can be applied to maximize an organization's operational efficiency and to provide a strategic, competitive edge to *your* business.

Many organizations have realized that potential for improvement exists through a Total Quality Management approach. However, many executives are dismayed by the jargon which surrounds the subject and are bewildered by the array of tools and techniques available.

To assist in overcoming these barriers, Part Three, the Executive Encyclopaedia, explains the essential jargon and provides technique briefings on the major improvement tools and techniques. The briefings show how these tools and techniques are applicable to all types and sizes of organization. When key terms listed in the Executive Encyclopaedia appear in the text, bold type is used to draw your attention to them.

Most importantly, Part Two guides readers in formulating their Total Quality Management approach and on selecting appropriate tools and techniques. These strategic decisions are critical, particularly as the range of techniques that can be applied is very wide. The selection process, however, is by no means straightforward. The report addresses this by providing guidance on how to review the needs of your own organization and how to make an informed selection of an appropriate mix of 'tools of change' to suit your business.

This report is intended for use as a manual and as a reference, so there is a bibliography and information on further sources of help at the back of the book.

## How to use this book

This report is designed both for casual readers and for organizational strategy makers. It is suggested that readers wishing to plan, in depth, their own improvement initiative should work sequentially through the book.

However, those wishing to browse for increased general awareness and those seeking specific topics should first concentrate on identifying their aims:

- Browsing to find out how to obtain commitment to change? In Part Two, read Section 2 – A model management framework for Total Quality, and Section 3 – Starting the change to Total Quality. Follow this with a study of Part Four, Profiles of Success Through Total Quality.
- Browsing to understand the jargon of Total Quality? Turn to Part Three, scan the contents list for the Executive Encyclopaedia and read the technique briefings of your choice.
- Needing information on implementing a specific technique? In Part Three, consult the contents list for the Executive Encyclopaedia and read the appropriate briefing. For further depth, refer to the bibliography at the back of the book.
- Interested to see what others have achieved by Total Quality? Read Part Four, Profiles of Success Through Total Quality.

Whatever your motives, you will not have obtained the full benefit from the time you spend with this book unless you complete your own version of the six elements of the management framework from Section 2 of Part Two.

We urge you to read on and, as an individual or with a team of colleagues from your organization, form your ideas for the charts in Figures 9 to 14 inclusive. Even if you do nothing else with this book, those ideas will be of great value in directing thinking towards Total Quality in your organization.


# Total Quality . . . is it really a strategic issue?

Successful organizations are making major changes in their business performance and customer orientation as a result of ever-changing global market-places. To highlight these strategic issues we asked Cathy Kramer, executive vice-president of the American Association for Quality and Participation, to give us her opinion on how American organizations are facing up to such changes.

First, we asked her, 'Where do American companies see the biggest threats and opportunities coming from?'

She replied, 'Since we all compete in a global market-place at this point, all countries have the ability to represent both threats and opportunities to the US. Certainly Europe post-1992 and the Pacific rim countries, including Japan, could be seen alternatively as enormous areas of opportunity for selling

American products and services or as potential threats to the American market. Within the US, there are opportunities in improving our performance and customer orientation in both industry and government.'

We then asked her to describe the American response through a Total Quality approach. Here is her reply.

'It has been almost forty years since Drs **Deming** and **Juran** carried their messages and lessons about quality methodology to Japan, a country which has grown to become our most significant competitor in the global market-place. Americans are weary of hearing of the Japanese successes and subsequently the threats they pose to our economy. But perhaps the weariness comes from not wanting to admit to ourselves that we made a serious tactical error in turning a deaf ear to Deming and Juran, an error that I believe we are working earnestly to correct.

'Even though it took most American companies thirty years before they began to take the foreign competition seriously, significant progress has been made in the last decade. The 1980s saw an explosion in the use of statistical methods and quality methodology in both manufacturing and non-manufacturing organizations. Our resistance to accept that a foreign competitor could be producing better products manifested itself in a "not invented here" syndrome. This caused us to drag our feet again in adapting the quality concepts to non-manufacturing environments. However, the second half of the 1980s can be characterized by the growing interest in quality improvement in service industries and government operations at the local, state and federal levels.

'Our slow start in applying quality techniques has been replaced with an impatience to see results and an increased energy for improvement in our major industries. We have moved from simply measuring productivity improvement to focusing on the customer, the customer's requirements and the customer's satisfaction. The early pioneers in American business are leading the dialogue about the importance of meeting the customer's requirements and these leaders are being heard, followed and held in high regard as examples of how it can work in this country too.

'The increased energy expended on and emphasis given to quality improvements is not only being felt by upper management in organizations but is being demonstrated daily by large segments of the American work-force, both union and non-union alike. And this is perhaps the most promising change offering the greatest optimism for our future. The fact that the words "co-operation, involvement and joint responsibility" are frequently found in labour management agreements is no small accomplishment in itself. The American work-force has finally collectively accepted the challenge of our foreign competitors and is now organizing toward the common goal.

'The techniques applied are numerous. Fortunately, we are in a learning mode and are willing to select from a host of possibilities those techniques that are appropriate to the business or organization. The variations on employee involvement are as numerous as the organizations applying them. Total Quality Management, statistical process controls, self-managing teams and redesign workplaces are just a few of the techniques now gaining ground in today's organizations.'

We thank Cathy for her contribution. We couldn't have made the point better. Hopefully your organization is not suffering from the 'deaf ear' or 'not invented here' syndrome. Whether you view the Single European Market (SEM) as an opportunity or a threat, it is a potent incentive to justify a Total Quality Management approach in your organization. This book will help you to find your way in and around the post-1992 world, provided that you can release the potential in your organization. If you have doubts, turn now to the profiles of success and scan the benefits achieved by just a sample of UK and multinational industries.

# Part Two
# A Total Quality Approach

# Overview

The objective of this part is to assist you in defining your own improvement strategies. To achieve this, a three-step approach is used.

Step 1   Involves understanding the six basic *concepts* underlying all Total Quality successes.

Step 2   Addresses the six *management elements* that must be integrated into the practices and systems of your managers.

Step 3   Reviews and plans the six *stages* of converting from the existing situation to the launch of these new methods.

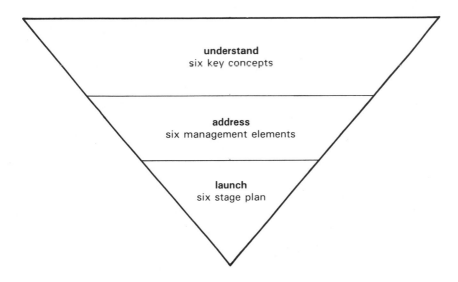

**understand**
six key concepts

**address**
six management elements

**launch**
six stage plan

**Figure 1**   Putting Total Quality into practice – three typical steps

Part Two examines each of these steps in turn.

First, section 1 describes the concepts and explains the questions that must be asked and answered to translate them into practice. Once the concepts are understood, they form a foundation to support the management framework required by a Total Quality organization.

Section 2 describes the elements of the management framework and identifies the considerations needed for each to be constructed. A sound framework provides secure links between concepts and practical application.

Finally, section 3 reviews the planning stages suggested to start the change process and begin the evolution of a Total Quality organization.

# Section 1 – The basic concepts of Total Quality

## Introduction – Six key concepts

As you progress through this part you will learn how to define and implement Total Quality processes into your business.

We do not contend that there is a single x-point strategy that is appropriate to any business. Instead, we will assist you to create your own. To do this, we will start by identifying the key concepts that have underpinned the success of those organizations which have fully realized the benefits of Total Quality.

It will then be possible to examine how these concepts can be developed into a management framework appropriate for your own organization.

The six key concepts to be considered are as follows:

1 Customers (external and internal)
2 Never-ending improvement
3 Control of business processes
4 'Upstream' preventive management
5 Ongoing preventive action
6 Leadership and teamwork

The following pages evaluate each in turn.

## Concept 1 – Customers (external and internal)

Successful organizations realize that a major factor in Total Quality improvement is monitoring performance in meeting or exceeding customer requirements.

In this context, they understand that 'customers' are not only the people to whom you sell products or offer services, but are also your internal staff. All your personnel interact with their own 'suppliers' in the processes they operate, not just the people who sell products or services to your organization. These internal suppliers 'down the line' in a process provide an input to the next job and thus satisfy their customers – the people 'up the line' who need the information or material.

For example, people machining components (process 1 in Figure 2) are customers of the material supplier, but are suppliers to the assembly shop (process 2 in Figure 2). Similarly, assemblers are customers of the component machinists, but are suppliers to the sales storemen. The storemen are suppliers to the external customer

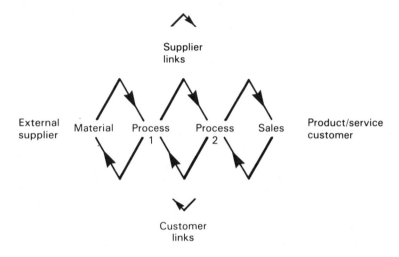

**Figure 2**    A chain of internal customers and suppliers proving continuous links between external supplier and external customer.

This concept – of a chain of suppliers meeting or exceeding customer requirements – is vital in the application of Total Quality. It demonstrates the responsibility that everyone in the organization has to contribute to quality by satisfying their internal customer so that, ultimately, the external customer is satisfied.

If each customer specifies and demands appropriate requirements, suppliers can focus their efforts to build quality into the process at each stage. Improvements are also more readily identified and implemented when a complex process is analysed in this way.

Acceptance of this concept into your organization raises several questions:

- How do you measure external customer satisfaction?

- How do you compare yourself (**benchmark**) with competitors?

- How do you identify and agree internal customer requirements?

- How do you display your current performance on customer satisfaction?

- Do your staff understand and accept the concept of internal customers?

## Concept 2 – Never-ending improvement

Attainment of world-class goals is only possible by striving for never-ending improvement in all aspects of performance. Improvement is a process that must never stop. Once targets are met, new ones are set, aiming for even higher levels of product, process and service efficiency.

In this way, a real competitive edge can develop by steadily widening the advantage over static or slow-changing competitors. The enviable reputation of Japan's industry has evolved largely because it has been enthusiastically operating this concept for forty years; competition has improved slowly, so the gap has increased to a significant lead.

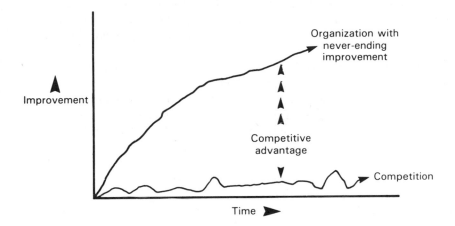

**Figure 3**   Never-ending improvement – the way to a competitive advantage.

Adopting this concept of never-ending improvement – and the other concepts of Total Quality – will involve changing your organization's management style. Total Quality cannot be just a 'bolt-on' programme of change. Your techniques must be targeted at steady, ongoing progress, not on short-term gains which may not be sustained. Measures of performance and management horizons will have to be adjusted to recognize the need for sustained improvement.

Implementation of strategies embracing the concept of never-ending improvement will raise issues such as:

- How do you maintain impetus to keep seeking further improvements?

- Which measurement and review processes should you use?

- How do you convince your staff that business and organizational survival will only occur if you achieve the steady and continuous improvement of everything they do?

# Concept 3 – Control of business processes

The quality of any organization's products or services is determined by the basic business or manufacturing *processes* that create them. If the chain of processes is made efficient and effective, then the resulting products or services will also be efficient and effective, and will satisfy the external customer.

Effort must therefore be directed to controlling the processes, rather than concentrating just on direct, specific controls of products or services. Applying direct product and service controls, such as inspection, often only addresses symptoms of potential problems, neglecting causes which lie within the process itself.

It is important to identify the process *owners* when reviewing process control. These are all the people who influence and control the process on a minute-by-minute, daily basis and who are therefore well qualified to advise and comment. The effective use of their knowledge and abilities is essential to effective process control and is a cornerstone of Total Quality. Leadership and teamwork is discussed later, as Concept 6.

Processes are many and varied, but each is important and should be controlled in appropriate ways. Manufacturing and service delivery processes are usually easy to identify, but less obvious supporting processes must not be ignored. Administration, secretarial and personal services such as typing, greeting visitors, receiving telephone calls, presenting invoices and so on must also be controlled because they are all meeting customer and business requirements.

Interpreting the concept of process control into an action plan will raise some vital questions:

- Have you identified the processes that are critical to your business?

- Do your staff realize that all work is a process, converting a range of inputs into a number of agreed outputs?

- Do you have staff who are clearly responsible for each process, irrespective of departmental allegiances?

Process analysis chart

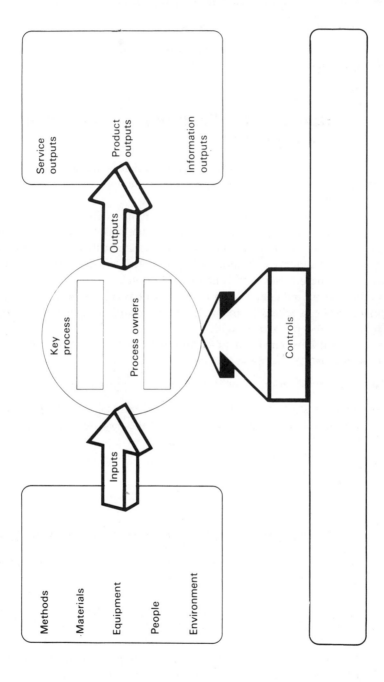

**Figure 4** The examination and analysis of a process to identify key elements.

- Who plans, controls and monitors each process?

- Do your staff feel able to criticize each process?

- Can your staff complete the headings in Figure 4 for their own key processes?

## Concept 4 – 'Upstream' preventive management

To the successful organization, improvement is about seeking out potential problems (or 'improvement opportunities') and *not* about waiting for a failure to identify an area for improvement.

'Upstream' management shifts the emphasis from past-event inspection to pre-event planning and prevention activities. Thus, in Figure 5, the upward arrows represent ways of obtaining information that will help to anticipate potential problems; continuous improvement means using this information to move the emphasis of control to the left.

Inspection only deals with history, so having the best inspection department in the world will not make a successful company, but managing by planning and prevention will.

To shift your emphasis to Total Quality Management by prevention will demand answers to these questions:

- What information is available from your processes?

- Do your staff, especially the process owners, understand that effective use of all available information is the key to control and improvement?

- How can the information be presented and used to assist control and continuous improvement?

## Concept 5 – Ongoing preventive action

Successful organizations constantly attack the real root causes of problems, or potential problems, that hinder staff in doing their jobs.

Staff must feel able to highlight any such hindrance. Your management and staff combined must be capable to decide and implement appropriate corrective and preventive action. The action must remove or minimize the root causes of the problems and prevent their recurrence.

'Firefighting' – panic management of symptoms to deal with crises – will then be minimized; it is not part of a Total Quality approach.

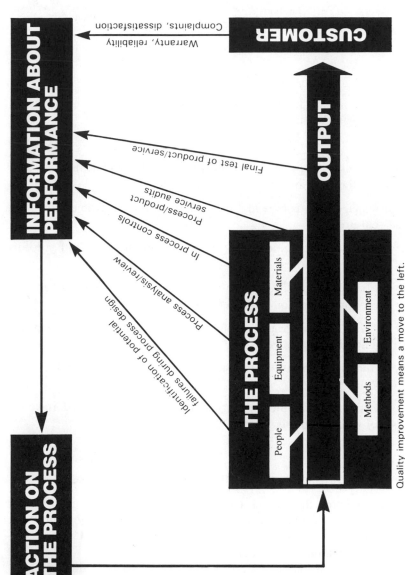

Quality improvement means a move to the left.

**Figure 5** 'Upstream' management by prevention depends on the use of all information available from the process.

**Break the loop**

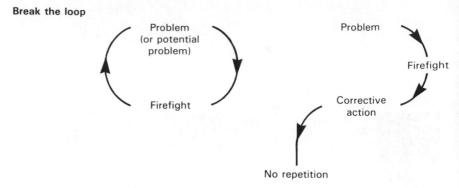

**Figure 6**  Break the loop of recurring problems by attacking root causes and eliminate the need to 'firefight'.

Incorporating this concept of ongoing preventive action will demand positive answers to further questions:

- Is your management team ready for open discussion on problems in a 'no blame' culture?

- Are you prepared to remove root causes of problems or will you just 'firefight' the symptoms?

- How seriously do you listen to staff, suppliers and customers when they identify opportunities for improvement?

- How do you manage, prioritize and co-ordinate such improvement and **problem-solving** activities?

- Do you maximize the use of all staff, at all levels, to the limit of their abilities?

- Have all your staff had the appropriate training required to undertake the necessary level of involvement?

## Concept 6 – Leadership and teamwork

Total Quality requires the highest standards of leadership and teamwork. This concept depends on participation and teamwork throughout the organization, in all activities.

Visible and genuine commitment from the chief executive is essential if the organization is to embark on a 'journey to excellence' through Total Quality.

Commitment from the management team is the only way to implement and maintain the culture necessary for Total Quality.

The sustained commitment of every person in the organization, whatever their role, is required to achieve improvement goals for Total Quality.

This motivation and commitment is achieved by extensive involvement of all the staff in the decision-making and problem-solving activities of business. Recognizing the individual's contribution and improving the organization's approach to teamwork can dramatically improve the return on your people costs – in remuneration, turnover and Total Quality terms.

In total, the benefits of Total Quality can only be fully realized by maximizing the use of all staff, at all levels, to the limit of their abilities.

Reviewing the application of the leadership and teamwork concept requires consideration of questions which are fundamental to the management style and philosophy of the chief executive and the organization:

- How do you communicate with your staff?

- How do you obtain their commitment and motivation?

- How good is your teamwork?

- Do you need to change your management style?

- Could you benefit from multilevel problem-solving/task teams?

## Summary

This section has identified the key concepts that underpin Total Quality success. The application of the six concepts includes recognizing customer needs, seeking never-ending improvements, controlling processes, managing 'upstream', taking ongoing preventive action and making the fullest use of people – through leadership and teamwork.

For full effect, the concepts must interact and support each other; no one concept should be applied in isolation.

Section 2 examines how these concepts can be developed into a management framework appropriate for your own organization. Before turning to the detail of section 2, it would be useful to read Part Four 'Profiles of Success Through Total Quality'. This illustrates how the basic concepts have been applied by organizations which are very different – but which are all very successful because of Total Quality.

**Figure 7** Full utilization of all your staff's abilities is essential.

# Section 2 – A model management framework for Total Quality

## Introduction – Six key elements

This model provides a guide to making decisions on your own improvement process and on how to manage and implement the necessary changes.

Before starting this section, make sure that you understand the six key *concepts* of Total Quality, that is:

1 customers (external and internal)
2 never-ending improvement
3 control of business processes
4 'upstream' preventive management
5 ongoing preventive action
6 leadership and teamwork.

Planning the improvement process for your organization is a very important matter. To start, it is necessary to define a series of strategies and methods for the key elements of a management framework by answering six questions:

1 What are the missions, aims and objectives that you wish to achieve in the short and long term and how are these to be communicated to staff, suppliers and customers?

2 How are you going to collate external customer and competitor intelligence?

3 How are you going to measure performance?

4 How are you going to highlight and communicate improvement opportunities?

5 By what means are improvement opportunities to be implemented?

6 How are you going to co-ordinate the Total Quality programme?

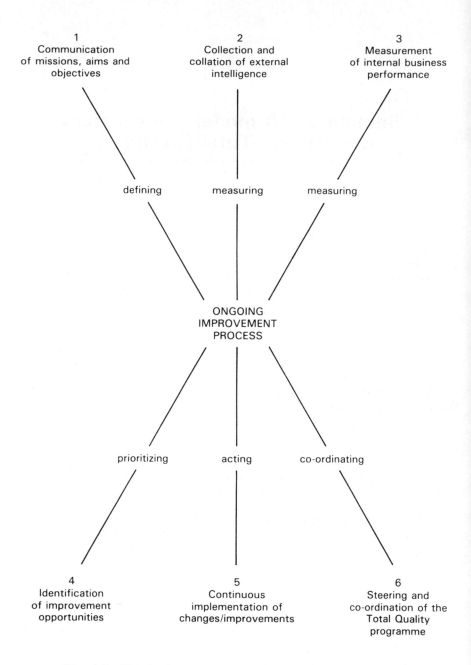

**Figure 8**    The six elements of a model management framework for Total Quality.

Figure 8 indicates how the answers to these questions integrate into an overall plan and addresses elements of the management framework.

On the following pages, each element of the management framework is considered in turn. A structure diagram is shown for each element; this can be copied and used to develop your own ideas by **brainstorming**.

In many organizations, these planning activities will be performed by a senior management or executive team and the structure diagrams will be used as aids to brainstorming in these groups.

When initial ideas for each of the elements have been generated, they should be checked to ensure that the six key concepts have been covered and in order to stimulate further ideas. Several rounds of brainstorming, drafting and redrafting will be needed before the next step – the implementation plan outlined in section 3 – can be considered.

# Element 1 – Communication of missions, aims and objectives

## Total Quality policy

Every organization needs to develop its own policies but at some time a definition of what Total Quality Management means to your organization, and each individual in it, is essential. Such a mission statement may be anywhere within a range from a simple sentence such as 'Meeting or exceeding the needs of customers at a price that presents value to them' (attributed to H. Harrington, an American quality expert) to a tailored version of more elaborative series of paragraphs such as the version, attributed to the British Quality Association, shown below:

> Total Quality Management (TQM) is a corporate business management philosophy which recognises that customer needs and business goals are inseparable. It is applicable within both industry and commerce.
>
> It ensures maximum effectiveness and efficiency within a business and secures commercial leadership by putting in place process and systems which will promote excellence, prevent errors and ensure that every aspect of the business is aligned to customer needs and the advancement of business goals without duplication or waste of effort by releasing the full potential of all employees.
>
> The commitment of TQM originates at the chief executive level in business and is promoted in all human activities. The accomplishment of quality is thus achieved by personal involvement and accountability, devoted to a continuous improvement process, with measurable levels of performance by all concerned.
>
> It involves every department, function and process in a business and the active commitment of all employees to meeting customer needs. In this regard the 'customers' of each employee are separately and individually identified.

Companies successful in this implementation of Total Quality will typically have statements embracing:

- why the organization exists
- what business the organization should or will be involved in
- what unique or distinctive competence the organization should concentrate on
- how the organization will conduct its business (its values).

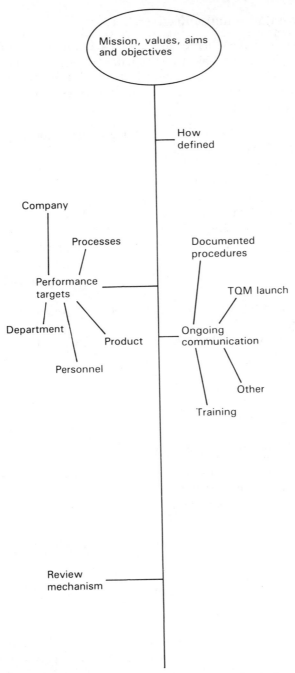

**Figure 9** Structure for your communication of missions, aims and objectives.

## Business aims and objectives

Once the aims of Total Quality have been defined, every organization will set itself a series of objectives that they intend to achieve through TQM. These business objectives are effectively the benefits to be gained and, although they will vary from organization to organization, often realized benefits include:

- the improvement of profitability by increased operational efficiency
- cultural and behavioural change
- the prevention of waste
- the improvement of customer satisfaction
- maintaining or increasing market share
- the achievement of product and business excellence
- releasing your organization's people potential
- the improving of product or service quality, product safety and reliability
- the minimization of loss to the individual, the company and the community
- associated improvements in operational safety, occupational health and the environment
- encouragement of each individual's personal improvement, innovation and creativity.

## Targets and measurements

Once set, measured progress to and achievement of these objectives may well become the overall performance measure of the success of your business, let alone its Total Quality strategy.

## Communication and training

Finally, note that:

- These aims and objectives will affect your staff at all levels. At the level of the individual or a process they may well result in specific performance targets.

- You will need to address methods of ongoing communication and review.

- Many of these aims and objectives for Total Quality will be new concepts to many of your staff and a programme of ongoing training will be needed.

Now, work through the detail in Figure 9 and add your own ideas.

# Element 2 – Collection of external intelligence

### The market

It can be said that the success of any organization is dependent on how accurately it has understood and defined the needs and expectations of its customers: first its external customers, who are essential for its continued existence, and then its internal customers, or employees; and on how efficiently it has then converted those defined needs, requirements and expectations into products or services which fully satisfy all parties.

### Customer attitudes and needs

The definition of an organization's specific Total Quality strategy will require answers to several questions about levels of customer satisfaction, both internal and external, and about both the perceived achievements and the costs of such performance. Whilst some organizations will have collected data capable of providing such an evaluation, others will need to initiate a series of market or business reviews.

The Executive Encyclopaedia provides more guidance on monitoring **customer satisfaction**.

### Comparisons with competitors

Many organizations will create business targets based on an evaluation of their competitors' performance. These measures provide an excellent internal focus for improvement and are often called 'competitor **benchmarks**'.

### The business environment

Environmental and societal factors are, of course, significant in these external evaluations and in many organizations the concepts of loss and **quality costs** are applied to society and not just to the business.

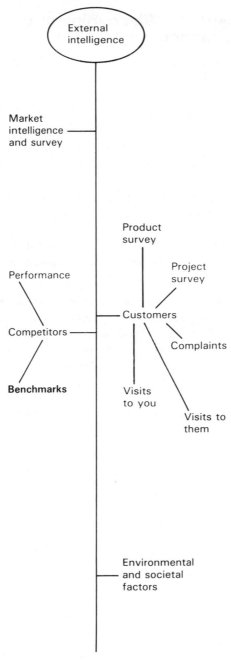

**Figure 10**  Structure for your collection and collation of external intelligence.

## Element 3 – Measurement of internal performance

Historically, many organizations have attempted to improve product quality solely by inspection at the end of a process, and by utilizing an independent team to perform it. Development of this approach is not the route to improvement. When evaluating your measurement strategies it is important to remember:

- First that Concept 3 – Control of business processes – will always lead you to suggest early application of process controls as opposed to late application of sorting.
- Secondly, Concept 6 – Leadership and teamwork – will make you utilize the individual performing the task, rather than an independent person, wherever possible.
- Finally, that measurements apply to all departments at all levels from the main board to the switchboard; i.e. their use is not just a technique for the manufacturing or service delivery areas.

Many and varied techniques are used, ranging from quality costs for overall focus and monitoring to statistical techniques for specific processes. At a workplace level, the techniques of **process management** and **performance measurement** are often applied to enable those responsible for the operation to understand their role fully and to be involved in the setting of challenging targets for their performance.

Can you enter techniques by each of the prompts in Figure 11?

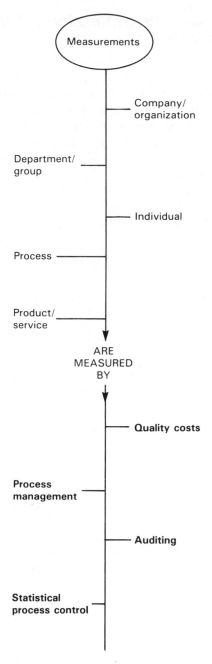

**Figure 11** Structure for your measurement of internal business performance.

## Element 4 – Identification of improvement opportunities

Successful organizations allocate responsibility for improvement both vertically, within an organizational structure, and horizontally, in the processes that flow across organizational boundaries. Many organizations find they are weakest in the horizontal dimension and apply effort to ensure cross-boundary, interdepartmental preventive actions are encouraged.

The Executive Encyclopedia (see E is for **Employee participation**) shows some typical methods used to achieve this.

Many organizations find the need for a formal system of communication for the improvement opportunities that cannot be managed within one individual's work group, and use documented systems of communication to highlight the need for corrective action.

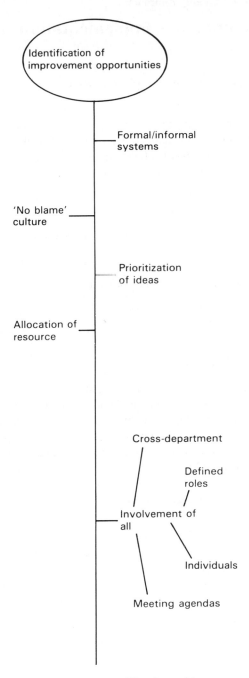

**Figure 12** Structure for your identification of improvement opportunities.

# Element 5 – Implementation of changes

## Culture and management style

Success is very often directly related to an organization's ability to create an environment that empowers and challenges its staff to change and improve their performance continually. If you hesitate with your answers to any of the questions below, you probably have a need to review your own management team's style and actions and to ask if your approach to teamworking is right. For example:

- Is firefighting still, secretly, a favourite, satisfying pastime?

- Are you trying to control people through systems?

- Do you have open communication and teamwork?

- How compartmentalized is your business?

- Do people identify with processes?

## Use of teams

One way that contributes to achieving such change is to manage your major business improvement projects by the use of teams trained in **problem-solving** techniques, thus promoting teamwork and involvement. However, a common stumbling-block to the implementation of this approach by 'bottom-up' methods is the reluctance of your existing management structure to accept these techniques, and you will need to implement a corporate drive to change leadership behaviour. It is not uncommon to use psychometric tests to identify managerial style inadequacies and subsequently to deal with them positively.

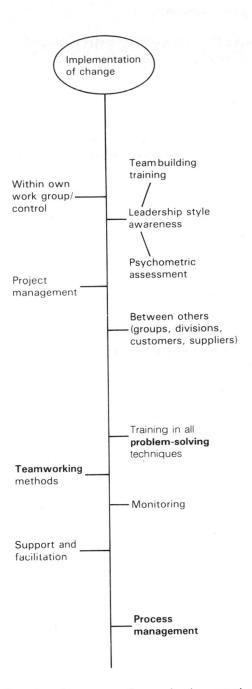

**Figure 13** Structure for your continuous implementation of changes/ improvements.

# Element 6 – Steering and co-ordination of the Total Quality programme

## Steering group

Although a steering group is found to be essential in many successful organizations, it is important that the responsibilities of each and every individual are accepted.

However, key issues such as resource prioritization, recognition of effort and performance updates/communication are often handled by a group to which is allocated responsibility for steering and co-ordination.

## Communication

It is important to communicate progress through suitable channels. You may already have a company newsletter or hold regular team briefings. Consider also the use of notices and personal communications.

See also E is for **Employee participation** and C is for **Communication** in the Executive Encyclopaedia.

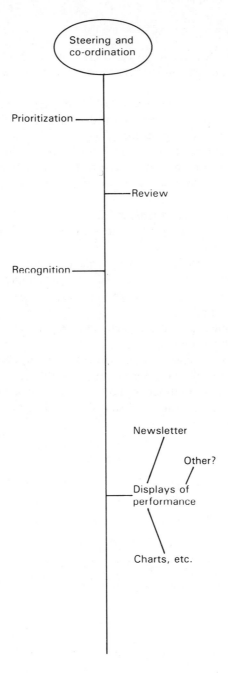

**Figure 14** Structure for your steering and co-ordination of the Total Quality programme.

## Summary

This section should have assisted you to review the needs of your organization and to design a management framework for your own improvement process.

Your ideas for each of the elements should be checked to ensure that the six key concepts have been covered and to stimulate further ideas until, after several rounds of brainstorming, drafting and redrafting, the next step – the implementation plan outlined in section 3 – can be considered.

However, don't leap into implementation without reviewing your commitments. Before you go further, ask yourself if you are prepared to dedicate yourself to the process.

A Total Quality style of management requires the chief executive to make time freely available for leading the change process and for demonstrating commitment through practical actions.

The changes in style often necessary to achieve a positive operating culture must be accepted by all other members of the management team.

Accountability for the achievement of key targets must be agreed so that progress can be measured.

If you are not completely confident that your preparations will obtain such commitment, strengthen your intended actions before launching the drive towards Total Quality. Going too soon will frustrate your efforts and might negate many of the benefits you are hoping to achieve.

Use the six-stage plan outlined in section 3 to check that your launch programme is comprehensive and will start the change process successfully.

# Section 3 – Starting the change to Total Quality

## Introduction – Six basic stages

You now have an understanding of the six key concepts of Total Quality and have formulated your thoughts about the ongoing needs for the six elements of a new management framework.

The initiation of these changes is an evolutionary process, requiring the following six stages:

Stage 1 – Statement of intent

Stage 2 – Awareness

Stage 3 – Diagnosis

Stage 4 – Initial strategy

Stage 5 – Management consensus

Stage 6 – Launch.

## Stage 1 – Statement of intent

An executive statement of the intention to evaluate the possible benefits of a new approach to business – a Total Quality approach – is a necessary preliminary to the change process.

## Stage 2 – Awareness

To begin the process of gaining commitment to the change, directors, executives and senior managers must be trained in the concepts, tools and techniques of Total Quality.

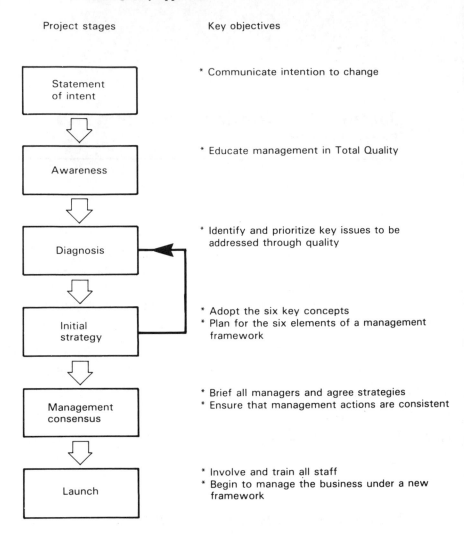

Project stages                    Key objectives

Statement
of intent

* Communicate intention to change

Awareness

* Educate management in Total Quality

Diagnosis

* Identify and prioritize key issues to be
  addressed through quality

Initial
strategy

* Adopt the six key concepts
* Plan for the six elements of a management
  framework

Management
consensus

* Brief all managers and agree strategies
* Ensure that management actions are consistent

Launch

* Involve and train all staff
* Begin to manage the business under a new
  framework

**Figure 15**  Evolution of a Total Quality approach through six stages.

This awareness can be achieved through a combination of various measures such as internal and external training programmes, reading, seminars or visits to organizations already practising Total Quality.

## Stage 3 – Diagnosis

The identification and quantification of possible benefits from the new (Total

Quality) approach to business should be carried out by individuals or teams drawn from senior managers.

A series of individual reviews should be undertaken to cover the main strategic concerns of the senior management, such as:

- external customer satisfaction (by a survey or interviews with the most significant customers)

- staff attitudes and morale (by interviews and evaluation of staff turnover)

- internal and external communications

- costs of current performance

- market standing related to competition

- efficiency of basic business processes (check a sample of orders or contracts to assess the controls and judge the likely magnitude of quality 'failure' cost)

- interfaces with suppliers.

## Stage 4 – Initial strategy

The diagnosis of Stage 3 will indicate to the chief executive and his senior management team:

- what the organization does best (and what it does worst!)
- major improvement opportunities (in priority order)
- initial bases of performance from which to measure benefits
- a basis for planning the way forward.

Specific strategies can now be developed for each arm of the management framework, commencing with a clear mission statement.

## Stage 5 – Management consensus

A major senior executive and management forum to evaluate and accept the quantification of the need to change (from Stage 3) and the proposals for your own strategy and management framework (Stage 4). This could be a straightforward board meeting (suitably extended in duration) or a dedicated, off-site conference.

## Stage 6 – Launch

A training and communication cascade down to achieve management and staff commitment throughout the organization. This must ultimately establish an awareness in each employee of Total Quality concepts, with a common language and 'jargon', and of your own aims, objectives, strategy and management framework.

This very significant activity requires a carefully planned training and communication framework. As for senior management awareness (Stage 2), this can be achieved by a combination of measures, including analysis and practical application in the workplace.

## Summary

By working through Sections 1 and 2, you will have come to understand the concepts of Total Quality and you will have applied these to develop a suitable management framework for your organization.

Section 3 has demanded a close examination of your organization, its performance and its operating environment. This will have quantified and prioritized all the major factors in your situation – those you knew intuitively and perhaps a few that you had not previously considered.

Armed with this information, you now have a strategy for starting the change to Total Quality. The strategy addresses the needs of your customers, accounts for the strengths and weaknesses of your organization and faces the threats and opportunities presented by your operating environment.

Your management and staff understand the need for change and are committed to it. You must now lead and sustain the progress towards Total Quality Management. We wish you well.

# Part Three

# Executive Encyclopaedia of Total Quality Processes and Techniques

Success in implementing change under the banner of Total Quality results from the careful choice of strategies and the selection of tools and techniques of change appropriate to your business.

This part contains an alphabetical overview of the most commonly applied Total Quality processes, explains some of the more commonly used acronyms and introduces some of the key players and organizations of world-class renown.

Entries are arranged alphabetically. There is a list of contents, and a full index appears later in the book. Terms which are printed in bold are ones for which there is a separate entry.

Readers should use this part to assist them in clarifying their thoughts on the possible application to their business of particular techniques.

Anyone wishing for in-depth advice on how to implement any of the techniques described is advised to support the information contained in these briefings with supplementary texts drawn from the bibliography.

Readers wishing to understand, in depth, the complete spectrum of quality vocabulary should also read BS4778: Part 1 (ISO8402), *Quality Vocabulary International Terms*, and BS4778: Part 2, *UK National Terms*.

# Contents

# A is for Attitudes

An important measure of the organization's performance is the attitudes of the workforce. It is particularly useful to know about these when starting the change process. A lot can be gained by sampling opinions on the need for quality improvement, the commitment of management, the level of knowledge of customers and suppliers and collecting any other general comments people may have in relation to the organization's quality performance.

Once the change process is under way, opinions may be asked again to measure improvements that have taken place in the eyes of the workforce.

It is helpful to use questionnaires to obtain this information. A typical example is given in Table 1.

*Table 1.* Example of an attitude questionnaire (for use internally within your organization)

---

**Quality improvement – the need to change**
The purpose of this survey is to assess openly the opinions held by all personnel, so as to ensure that the subsequent TQM programme takes account of each individual's responses. The questionnaire is being given to, and responses are required from, every employee.

Your name is requested so that XYZ staff can discuss a sample of the responses with individual people. This information will remain confidential to XYZ: i.e., other than to XYZ staff, *no* single individual will be identified with their responses.

1   What role will Total Quality play in strengthening your organization's ability to compete in a highly competitive market-place?
Very significant
Significant
Some
Not much
None

2   To what extent has your overall quality performance improved over the last two years (or over the period of time you have been employed, if less than two years)?
Very significantly
Significantly
Some
Not much
Declined

3   List the most important product or customer service feature(s) that you feel your customer would regard as being the most quality critical.

4   How committed are the management team to improving quality and performance?
Very enthusiastic
Enthusiastic
Committed
Partially committed
Not at all committed

5   Who are your internal customer(s)?

6   Which departments are your internal suppliers?

7 How do you rate your suppliers' performance?
Very good
Good
Average
Poor
Very poor

8 How much of your time is spent investigating or rectifying other people's preventable problems?
0%
25%
50%
75%
100%

9 What would you suggest is your department's key quality performance measure? Name one only.

10 What kind of role can you, in your current position, have in improving overall quality and performance?
Very significant
Significant
Some
Not much
None

11 List the single most important thing that you feel should be done to improve performance within your work area.

12 Any other comment you may wish to make?

_____

_____

Name _____

# A is for Auditing

See also M is for Management systems.

## Quality audits

Quality auditing is being viewed by many companies as a powerful tool in managing the business, particularly as it provides a mechanism for the identification of improvement opportunities. Auditing techniques can be applied at a number of levels:

- the quality system audit to assess quality management systems and styles (this can be conducted by the organization or by second and third parties)
- the product audit to evaluate the conformance of product to specification
- the process audit to monitor the effectiveness of specific operations
- the self-audit to review individual performance.

In the UK this activity has traditionally been in the domain of the quality department and carried out by them, very much in the role of policemen. Recent trends have brought about the need for independent audits, and many are seeing this as an opportunity to involve more of their management team.

In introducing the concept of the internal customer, the auditing of a department's internal suppliers can be a useful forum for contact, discussion and feedback. However, auditing must be carried out with the aim of finding opportunities for corrective action and quality improvement, not for casting blame.

The widespread adoption of **ISO9000/BS5750/EN29000** has greatly affected audit activities within the UK. The ISO9000 standard has a specific requirement that the company conducts a series of planned internal audits of the quality system and therefore the importance and frequencies of application of these internal audits has greatly increased.

To support these auditing requirements the British Standards Institution and the International Standards Organization have issued Standards BS7229: 1991 and ISO10011: Part 1: 1989 respectively. BS7229 also includes guidance on the quality of quality systems auditors and on managing audit programmes.

Whilst further guidance can be found in the standards mentioned above, the success of the techniques of auditing is predetermined by the way they are introduced into any business culture. They must be seen as providing a 'no blame' forum for review, discussion and improvement. The following are some key, and often overlooked, factors for consideration by any audit team.

### Discussion not interrogation

The emphasis in an audit should be on discussion with departmental personnel on how they carry out their activities and about any problems they experience. There should not be an outside observer standing remote from the activities and scrutinizing them, nor just a question-and-answer quiz.

### Listen

Initiate discussion by pertinent questions, but then adopt the role of listener. The more talking you do, the less you are going to find out.

### Don't criticize or argue

It is not the role of the auditor to pass judgement on an individual's activities. Auditing is carried out to determine objectively if the quality system is being operated as documented. Any element of perceived personal criticism or outright argument is counterproductive to obtaining the necessary objective evidence of compliance to the system.

### Verify non-compliances

Where you discover a discrepancy, establish a non-compliance based on facts, not on subjective opinion. Where possible, verify from records. Determine whether the necessary records exist; don't record a non-compliance just because records have not been automatically presented to you.

### Be specific

When a non-compliance is found, the proposed corrective action should be discussed with either the nominated escort or the head of department. The department should be given the opportunity to recommend the appropriate corrective action and give advice if requested. The corrective action should be agreed between the auditor and the department and recorded. The time-scale for following up the corrective action should be agreed and recorded.

### Seek clarification

While the knowledge of the auditor, with respect to the quality system standard and the relevant part of the quality system, is important, he/she cannot be expected to know everything. If any doubt arises either in defining a non-compliance or suitable corrective action, then clarification should be sought from the management representative for the quality system. Don't resort to guesses or subjective opinion, or, worse, ignore a potential problem.

### Don't be secretive

In the course of an audit it is necessary to make notes on observations and comments which may need to be followed up by checking records. The open discussion necessary for a good audit will not be achieved if furtive note-taking is perceived by the department. Adopt a stance which will allow the escort and, if possible, the individual involved to see the information being recorded.

### Give credit where due

The personnel you have involved may be operating under difficult conditions and the audit may only be adding to the pressure. If they are operating the system effectively, then ensure that message is conveyed to them. If they have found a way of improving the system, then, though it may result in non-compliance to the current system, concentrate on agreeing the necessary action to amend the system rather than emphasizing the non-compliance.

### Control time

Although it is important to listen to departmental personnel, ensure that you are not side-tracked or subjected to all the grievances of the department, and that you do not allow the discussion to degenerate into gossip. From time to time it may be necessary to insert another pertinent question to keep the audit moving in the right direction.

## A is for Awards

External recognition by an independent, influential third party is often a powerful motivator for organizations' quality initiatives.

Certification to **BS5750/ISO9000/EN29000** (see M is for Management systems) has been the main award pursued by organizations in search of prestige and marketing advantage. Total Quality Management, however, is seen to be beyond the level of this certification and BS7850: Parts 1 and 2: 1992 have been published to extend the coverage of standards and, therefore, assessments.

In the USA, annual Malcolm Baldrige awards for excellence against a rigorous set of objective criteria are made to a handful of organizations. These awards are well established and highly prestigious.

The European Foundation for Quality Management (EFQM) has similarly developed searching objective criteria to assess the quality maturity of organizations. These criteria are used to judge entrants for EFQM's European Quality Company Awards.

The criteria used for these awards can be used by any organization for self-assessment of its level of absolute or relative quality maturity. Especially after certification to **BS5750/ISO9000/EN29000** has been achieved, these criteria and awards are increasingly popular.

## B is for Benchmarking

Benchmarking is the practice of measuring and comparing key aspects of your organization with those in other organizations to establish measures of relative performance, assist in setting new targets and discovering ideas for improvement.

The term is borrowed from land surveying, where a reference point established as a base for surveys is called a 'benchmark'.

Aspects to be compared include:

- costs and prices
- methods
- features
- service levels
- practices
- processes
- customer satisfaction.

Benchmarking establishes reference points – benchmarks – to describe what the market is getting from competitors and, in some cases, what the market expects in the future.

Methods include:

- opinion surveys
- examination/dismantling
- trial purchasing
- telephoned enquiries
- analysis of annual report and other published information
- following trade journals or trade association contacts
- using consultants.

It is obvious that competitively benchmarking aspects of your products and/or services will be useful, but non-competitor benchmarking also produces many benefits and is often easier to accomplish, because it can be done with the co-operation of the organization being benchmarked. Indeed, the mutual advantages of this approach often lead to a ready exchange of information and facilities, especially if the organizations are trading partners or parts (e.g. depots, sites or divisions) of a larger organization.

Study visits, exchange visits and other internal methods are used for non-competitor benchmarking. On the other hand, competitor benchmarking methods usually have to be external – customer surveys, sample buying, stripping and studying competitors' products and so on.

Ideally, benchmarking will achieve:

- a sense of proportion and perspective
- knowledge of best practices
- the setting of new, ambitious targets
- new interest for staff.

# B is for Brainstorming

Brainstorming is a simple technique that can be used to encourage group creativity. It is a formal approach used to help a group generate as many ideas as possible in as short a time as necessary on a chosen subject. Maximum benefit will be gained by adopting formal guidelines based upon three main features.

- Cross-fertilization – this happens when two or more people have part of an idea which on its own may seem irrelevant but when all are brought together a useful original idea is generated.
- Suspending judgement – this is crucial; the brainstorming session is purely for generation of ideas, not evaluation. Suspending judgement helps to avoid looking in on one particular area of ideas, thus exhausting opportunities to explore all the possibilities. No idea shold be considered ridiculous. It is part of the chairman's role to prevent participants from making comments such as, 'That would never work because . . . ' or 'We've done it before'.
- Formality of setting – particularly at the early stages, this helps to remove some of the tension that people may feel, which makes them hesitant to

suggest ideas. As people become more familiar with the technique and more used to expressing ideas, the setting has less of an influence.

In striving to meet these requirements, certain guidelines should be followed and it is the role of the chairman to ensure this happens. These rules are as follows:

- Define the central issue and make sure everyone agrees upon it.
- Everone should be allowed and encouraged to contribute; no one person should dominate.
- Every idea should be recorded in the words of the speaker.
- Never criticize ideas.
- Make no attempt to evaluate ideas.
- Don't develop ideas at length or get involved in lengthy discussions.
- The session should run for a set time or until all ideas have been exhausted, whichever is the shorter.

Great enjoyment and feeling of contributing can be gained from brainstorming, particularly when an idea is found that would not have come from an individual member of the team alone.

The underlying goal of brainstorming is the number of ideas generated, not the quality.

See also P is for Problem solving.

# B is for BS5750

The British Standards series (ISO9000/EN29000 series) is fully described under M is for Management systems.

# C is for Capability

### Capability studies

Capability is a measure of the *inherent* variation of a machine or process.

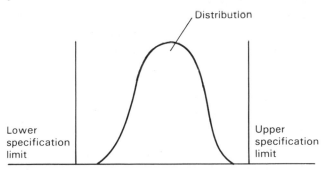

**Figure 16**   Frequency distribution of a 'capable' process.

The capability of a machine or process can be determined by conducting capability studies. These studies involve the analysis of a small sample of output to enable an estimate to be made of the overall variation likely to occur in a longer run.

In other words, capability studies are carried out to try to find out if a machine or process is capable of producing products or output that is within the required specification.

Capability studies are either done on a discrete operation with minimum external influences (machine capability study) or with a combination of influences from people, equipment, materials, methods and environment (process capability study). Machines are generally considered capable if four standard deviations are within the specification. Processes are normally considered capable if three standard deviations are within the specification. These minimum requirements and capabilities should continually be improved on where possible.

The capability of a machine must be better than product specification to allow for variation in other process variables; for example, materials, toolwear, machine setting and so on.

The results of a capability study provide a snapshot of the machine's capability at a point in time. As such they do not measure any variation which may occur with time. Therefore a second technique is used to take account of this element. The technique is ongoing process control using process control charts. Once control charts have been used to bring the process 'in control', the capability will be a closer reflection of the ongoing performance.

See also S is for Statistical process control and S is for Seven statistical tools.

## C is for Cause and effect diagram

A cause and effect (C & E) diagram is a tool which can be variously used to identify possible quality problem causes or as a next step to evaluate ideas generated at a **brainstorming** session.

The C & E diagram is also known as a fishbone diagram, because of its appearance, or an Ishikawa diagram after its founder. A typical diagram would appear as shown in Figure 17.

The problem or its effect, is stated on the right-hand side of the diagram and the more detailed it is, the better the result.

The various causes are sorted out and placed into categories; each category then comprises one main branch and each cause in that category forms a sub-branch off it. Furthermore, each sub-branch may then be divided into sub-causes and shown as further sub-branches.

There are various standard categories used, although there is no restriction. Typical headings may be:

- 5Ms – men, methods, machinery, material, maintenance.
- Major causes – major cause 1, major cause 2, major cause 3, major cause 4.

Various guidelines should be followed for successful C & E analysis:

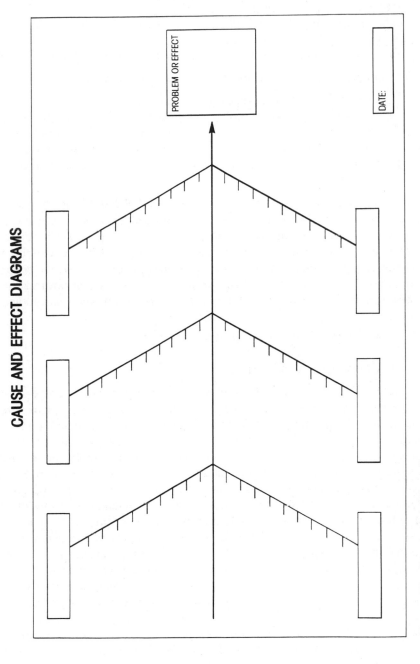

**Figure 17** Cause and effect diagram.

- Push causes as far back as possible.
- Deal with causes/influences, not symptoms.
- Make sure the problem definition is satisfactory.
- Use as many flow charts as possible on the causes.
- Don't let the diagram get too big if it appears to be in danger of breaking down.

The potential benefits of using C & E diagram analysis are as follows:

- It is a group tool: it encourages people to work together.
- It is invaluable as a tool to build a list of potential causes of problems.
- It can direct data collection to identify the true cause of a problem.
- The nature of the diagram leads people to think about variability.
- The diagram can be used as a management tool in allocating tasks for investigation.

The latest development of the C & E diagram is CEDAC, which is a cause and effect diagram with the addition of cards. Basically, this is an extension of the classical C & E diagram whereby the chart is displayed in the workplace and contribution to causes and solutions can be made at any time by the workforce. Progress on quality problems is instantly available and people have the opportunity to make a contribution whenever they discover relevant information.

See also P is for Problem solving and S is for Seven statistical tools.

## C is for Commitment

### Commitment to change

Commitment is a much abused word in the quality profession. Some see it as an excuse; others see it as an opportunity. One thing is certain: implementing Total Quality Management within an organization requires a recognition and an acceptance of the need to change. The extent of change will vary from company to company, depending on the existing culture. In any case, what is required is a deep commitment, at all levels, to making the change and seeing it through. For many individuals, the extent of change is difficult to accept unless they can see that the company genuinely intends to change course on a long-term basis. If the initiative is seen as another 'quick fix', then only lip-service will be given to the exercise by most levels of management. If the initiative is not to run aground on the rocks of cynicism, then a clear message of the course direction needs to come from the captain on the bridge.

The act of senior management issuing quality policy statements is not enough, and many organizaitons will take a long, hard look at their leadership style during the launch of a Total Quality initiative. It is not uncommon for a teambuilding programme, based often on a framework of psychometric analysis, to be initiated during the launch in order to develop staff teambulding and assist in the winning of commitment to change. Ultimately, example and demonstration that management take quality – and the staff – seriously are essential to win and maintain commitment.

# C is for Communication

Demonstrating commitment to quality improvement is very much about involvement. Utilizing opportunities for communication helps to demonstrate involvement. An important aspect of management leadership and style is the ability to provide and receive feedback. Therefore when you review the subject of communication, remember that it has to be two-way.

To successfully achieve this two-way communication, a forum for discussion and an infrastructure which will allow information to be channelled up and down your organization needs to be established. The forum should closely relate to the natural work groupings within the company, so that it can be put into the context of what people are trying to achieve by working together. The channels for communication need to be integrated with the normal routes by which people receive their instructions and have their priorities set. If people receive information on quality from one source but their programme of work from another, there is opportunity for conflict. This will certainly lead to confusion and is likely to result in demotivation, whereas consistency in communication will provide the much needed element of motivation to sustain any initiative in an organization. In particular, staff need to see that their views are being listened to. If their suggestions are not being implemented, they should have fed back to them an explanation of the reasons.

An important resource, in short supply in most companies, but essential for good communication, is time. Time needs to be invested in providing good communication; returns can be achieved by achieving quality requirements first time round and thereby reducing waste of time. To achieve the effective use of time, briefings should be properly structured with prepared notes so that the message required is clearly and consistently disseminated. Unresolved issues should be recorded for communication back-up to a level where a response can be obtained and fed back.

## Methods of communication

There are obviously situations in which processes cannot be stopped or telephones left unattended while briefings take place. In these instances, breaking natural work groups down into smaller groups can be done to overcome the limitations. Some groups, particularly in service-related organizations, can be geographically widespread. Here sometimes written communication may need to be substituted for face-to-face contact. However, two-way communication needs to be included in the process. Even when briefing sessions can be held, written communication in the form of quality noticeboards, quality bulletins and newsletters are sometimes used to augment the process. By providing greater background and in-depth detail they can reduce the time needed for briefing sessions. They are also a useful medium for publishing the successes achieved and recognizing the efforts of those involved.

Employees are always interested in knowing more about the business aspects of their workplace and organization, but a recent survey has shown that the sources for that information are not, in all cases, those that employees would wish them to have. It is perhaps worthwhile to consider the employees' priority list, showing the sources of information in order of preference, and relate it to your own workplace:

1   Immediate supervisor
2   Small group meetings
3   Top executives
4   Annual business report to employees
5   Employee handbook/booklets
6   Orientation programme
7   Local employee publication
8   General employee publication
9   Bulletin boards
10   Upward communication programmes
11   Mass meetings
12   Audio-visual programmes
13   Union
14   Grapevine
15   Mass media.

### Horizontal communication

So far we have looked at the process of vertical communication within an organization. An equally important aspect is horizontal communication between groups. We have discussed the concept of internal customers, but there are practical problems in every individual trying to search out and communicate with every individual internal customer. The concept becomes more practical when we recognize that individuals work in groups to serve the needs of internal customer groups. However, the practicalities of a whole group breaking off to go and discuss their needs with another group could be a limiting factor in horizontal communication. There is a need to establish a communication link which co-ordinates the activities of establishing internal customer needs and obtaining feedback on performance related to those needs. Effective organization can help greatly in this respect.

See also E is to Employee participation, O is for Organization and structure and V is for Visual control.

# C is for Control charts

Control charts are a statistical technique used to measure and reduce the variability of a process. They can be used as an ongoing measurement of various key parameters of a process so that action can be taken as soon as a problem is indicated. They can also be used as an analysis of process performance for **problem-solving** and improvement reasons.

Control charts are used to discriminate between the two types of variation that can be present in a process. These are:

- Common disturbances – variation that is inherent in the process.

- Special disturbances – the unpredictable variation that suggests a problem exists within the process.

If used properly, control charts can help to eliminate special disturbances, whereas common disturbances will only improve with a change in the nature of the process itself. This will usually be under management control.

Each control chart is initially produced by calculating what are known as control limits. Data are collected and then used as part of a statistical calculation to determine the control limits. A chart can then be drawn up as shown in Figure 18.

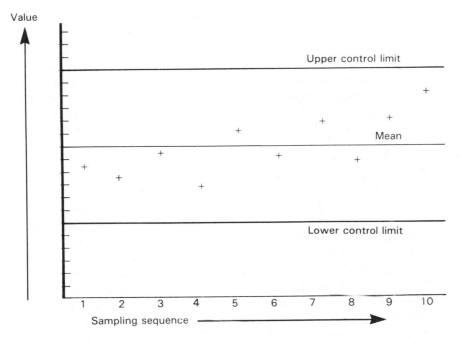

**Figure 18** Control chart showing a process parameter 'in control'.

Once this has been done, data can then be collected and entered as each sample is taken. If a sample lies outside the control limit, action is taken to eliminate the special cause responsible. It is important at this stage that the special cause is totally eliminated and steps taken so that it cannot occur again. When all sample data continue to fall between the control limits, the process can then be said to be 'in control'. At this point an assessment of the process capability can be made.

It is usual when using control charts to record data on two different types of charts. One type monitors the average of the sample taken; the other ranges between the highest and lowest value of the sample. These are known as mean and range charts.

See also S is for Seven statistical tools and S is for Statistical process control.

# C is for Corrective action

Corrective action is a key concept in planning your improvement activities and your ability to achieve it will be a prerequisite of success.

It is interesting to note that one of the main lessons companies learn from implementing basic quality management systems, such as **BS5750**, is the importance of corrective action. Companies can always, with varying degrees of success, detect defects and errors in their products and services, and then congratulate themselves on taking remedial action to put right those defects and errors. Very seldom do they seek out the cause of the problem and take corrective action to eliminate the cause. Hence they continue firefighting and rectifying the same problems week after week, month after month and, depressingly, year after year.

Corrective action means not just fixing the problem but removing the root cause of the problem so you don't have to fix it again. In other words, you need to break the loop, as shown in Figure 6.

Quality management systems, such as BS5750, result in a reporting system to ensure that the need for corrective action is highlighted and action is taken. The emphasis is very much on having a system which is carried out by management. However, corrective action is not just a system; it is also an attitude of mind and should become a way of life. To this extent it cannot remain the preserve of management; managers do not have time to solve all the problems, even if they do know all the answers. Therefore in delegating the responsibility for managing quality down through all levels, it is necessary to ensure there is an understanding of what real corrective action is. Ask these questions:

- Is corrective action understood at all levels in your organization?

- Have you an effective method?

- Do all your staff understand their responsibilities?

See also M is for Management systems.

# C is for Crosby

Philip B. Crosby Snr has obtained world recognition for his consultancy and approach to quality improvement.

After serving in the navy in the Korean War, Crosby became quality manager on the first Pershing missile programme. It was here that the concept of zero defects was born. He worked his way up through ITT and for fourteen years was corporate vice-president with worldwide responsibilities for quality. Following the success of his book *Quality is Free*, published in 1979, Crosby moved to Florida to set up Philip Crosby Associates and the Quality College which has 'enlightened' over 60,000 executives.

Crosby advocates a top-down approach; this is because, like other gurus, he believes that over 80% of problems are management caused and fewer than 20% are caused by workers. For this reason executives and managers are the first to be educated under a Crosby programme. They then cascade the concepts to all employees in their organization. The 'four absolutes' are the cornerstones of the Crosby philosophy. These are:

- Quality is defined as conformance to requirements. Quality does not mean goodness or excellence; we should all concentrate on identifying requirements and improving them.

- The system for causing quality is prevention, not appraisal.

- The performance standard must be zero defects, not 'that's close enough'.

- The measurement of quality is the price of non-conformance. Crosby estimates that manufacturing companies spend 25% of turnover on doing things wrong or reworking, and service companies spend up to 40% on non-conformance.

To improve quality (and to reduce the price of non-conformity) Crosby has proposed a fourteen-step approach. This is as follows:

1 Management commitment – to make it clear where management stand on quality. A policy statement should be agreed by operating management and implemented.

2 Quality improvement team – the team is set up to manage the quality improvement programme. Members of the team should be drawn from across the company.

3 Quality measurement – the method is to display current and potential non-conformance problems in a manner which can be readily understood and evaluated. The objective is to evaluate trends, identify problems and develop corrective actions.

4 Cost of quality – to define and measure the cost of quality. This is the sum of failure, appraisal and prevention costs, and according to Crosby it should be less than 2.5% of sales turnover.

5 Quality awareness – quality awareness should be a low-key activity and comprise two activities: regular management/employee communication meetings and published information through posters and house magazines.

6 Corrective action – as problems are identified by previous steps they should be tackled to prevent recurrence. Problems should be solved by local work groups if possible; if they are complex or long range they can be referred to a task team specially established by management.

7 Zero defect planning – zero defects (ZD) is a central plank in the strategy and therefore planning how to launch and how to gain employee commitment to this phase is important.

8 Employee education – training, particularly of supervisors, is essential if the ZD launch is to succeed.

9 ZD day – the ZD day is an event which ensures all employees know there has been a change. It is frequently accompanied by celebrations and show-business razzmatazz.

10 Goal setting – supervisors encourage their work groups to set improvement goals – e.g. reduction in defectives – and publish results of improvements.

11 Error cause removal – employees are encouraged to submit problems. Management has to react to these suggestions to maintain the momentum of the programme.

12 Recognition – all individuals should receive appreciation for their contribution; the most effective recognition is not financial but peer group oriented.

13 Quality councils – quality professionals should meet on a regular basis to develop themes for improvements.

14 Do it all over again – the quality improvement programme of never-ending improvement.

See also G is for Guru.

# C is for Customer satisfaction

## The dangers of not monitoring customer satisfaction

One view of Total Quality Management is to look at it as a loop which begins and ends with the customer.

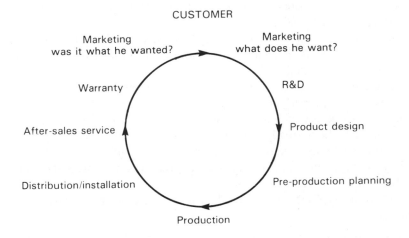

**Figure 19**    TQM – a new concept.

While many companies will expend considerable effort to determine what customers want before providing the product or service, often much less effort is devoted to closing the loop and finding out if they were satisfied. In many companies the measure of this success is taken as the level of customer complaints. An analogy which demonstrates the dangers in relying on this approach has been tagged Factor 42 and is attributed to the Ford Motor Company in America.

Surveys found that for each customer who made a complaint about their product (a particular car), there were, on average, six others with similar problems who hadn't complained. They also found that on average these seven dissatisfied customers would each tell six others and potentially dissuade them from buying the product. As a consequence, when one problem is highlighted there are in fact potentially forty-two problems. What do you think the factor is for your business?

## Methods of assessment

The assessment of customer satisfaction cannot be left to purely reactive techniques. Many organizations are now realizing the benefits of moving away from traditional feedback methods that have relied merely on negative responses – for example, a review of the complaints received – and are adopting new and pro-active tools of assessment as essential elements of their improvement strategies. Additionally they are realizing that the introduction of such new methods is often an opportunity to create a useful focal point from which to initiate other internal changes.

The techniques listed below are representative of the sort of pro-active methods from which an organization should select an appropriate mix and develop its own strategies.

It is important to note that an effective implementation of these tools can only occur if a properly resourced and planned strategy is initiated. The resources to support such a strategy should *not* be limited to revised roles for those staff who normally interface with the customer or assess the market. For example, the involvement of senior managers and the members of task teams adds to the importance of and the degree of commitment shown to the feedback you will receive.

Some of the typically used methods, which should be applied on an ongoing basis, include:

- surveying account customers by both formal questioning and informal group discussions
- sending senior managers to assess, by direct fact-to-face meetings, the degree of satisfaction achieved with key customers
- recording customers' views when they visit your site
- systematically recording customer views when your staff visit your customers' sites
- evaluating customer satisfaction during contracts and particularly on completion of transactions
- reviewing customer complaints
- reviewing service or field staff's reports
- holding focus group meetings with your own front-line customer interface staff
- issuing customer questionnaires with your product or service at the point of sale
- systematically reviewing and following up 'lost sales'
- utilizing the services of market research organizations.

Whatever methods are adopted, they should be capable of providing your management team with ongoing feedback and allowing your organization to quantify its success in terms of ongoing customer satisfaction improvement.

# D is for Data

The collation and analysis of data is key to the process of quality improvement. If you can't measure something, how do you know if there has been an improvement? On the other hand, it is important not to overdo it: the data must be useful, relevant, timely and accurate.

Data collection can specifically:

- help identify problem areas
- highlight basic causes
- establish monitoring procedures.

It is important to have an agreed reason for collecting data and a plan on how they will be analysed. Therefore before starting you should know the who, why, when, what, where and how of the data collection.

It is useful to recognize that there are two types of data. Hard data are those which are objective, typically facts and figures, whereas soft data are subjective and tend to be based on feelings and emotions.

Typical sources of hard data are shown in Table 2. They have been grouped under four common data categories.

Table 2. Typical sources of hard data

| People | Process | Product | Procedures |
|---|---|---|---|
| Training records | Plant records | Quality control | Maintenance |
| Safety records | Process control | records | Critical |
| Job descriptions | information | Customer | incident charts |
| Task descriptions | Unplanned | complaints | Work study |
|  | breakdown | **Statistical process** | measurement |
|  | Process flow | **control** information | Audits |
|  | analysis | Product standards |  |
|  | Process analysis |  |  |

Management information systems exist to provide managers with an overview of the operation and as such draw on different sources of information to present an overall performance picture, using only the relevant measures.

Typical soft data sources are listed below:

- group discussions
- interviews
- attitude surveys
- observations
- informal discussions.

Care must be taken when collecting soft data. People should not feel threatened; therefore the data must be confidential and non-attributable. It is useful to cross-check soft data, and it is useful to know how different people view the same subject.

# D is for Deming

Dr William Edwards Deming has achieved world-class recognition for his contributions to quality improvement. He was born in Iowa in October 1900. He received his PhD in mathematical physics from Yale University, then joined the US Department of Agriculture as a mathematical physicist. Deming studied under the eminent statistician R.A. Fisher in London, and he was also influenced by the originator of statistical process control '(SPC), Dr Walter Shewhart.

During 1941 Deming lectured in statistical methods to American industrialists, engineers and inspectors primarily involved in the war effort. The programme had a dramatic effect on productivity and scrap reduction. When the war was over, these advances were not sustained, which Deming attributed to failure to get the message through to management.

Following a visit to Japan, connected with a Japanese census, Deming was invited by the Japanese Union of Scientists and Engineers (JUSE) to present a lecture course to Japanese research workers, plant managers and engineers on quality control methods. Deming, learning from previous experience, requested that he should also give a presentation to Japan's chief executive officers. During 1950 he spoke to 100 of the industrial leaders; in 1951 he presented to a further 400. In the years following, the widespread adoption of his ideas became fundamental to the Japanese, whose industries developed to lead the world. In the USA, however, it was not until 1979 that Deming was listened to, when he was working with the Nashua Corporation. Deming has been awarded one of Japan's most distinguished honours, the Second Order of the Sacred Treasure, for his contribution to Japanese quality. The Deming Prize and Deming Price for Application are awarded by JUSE in honour of his contribution to quality.

Deming has identified the customer as 'the most important part of the production line. It will not suffice to have customers that are merely satisfied – customers that are delighted with your products and services will return again and bring new business with them'. It therefore follows that the supplier should develop products and services ahead of customer demands and not wait until the customer needs them because by then it will be too late and competitors will have stolen the market.

The Deming approach is the systematic improvement of quality by application of the Deming cycle (Figure 20). Improved quality leads to reduced rework, fewer delays and better utilization of equipment. As productivity improves and the company is able to market better quality at a lower price, the company is in a good position to provide security of employment and consequently stay in business.

Deming's lectures in the 1950s drew upon his statistical background, where one of the central themes was variability of the process. In each process, whether it be a manufacturing or service operation, there are two contributions to variability – those attributable to individual machines or operations ('special causes'), and those related to weakness of the management systems ('common causes'). It is to the elimination of the common causes that Deming addresses his 'fourteen points'. Deming saw the fourteen points as the basis for change:

1   Create constancy of purpose towards improvement of product and service, with the aim to become competitive, to stay in business and to provide jobs.
2   Adopt the new philosophy. We are in a new economic age, created by Japan. Transformation of Western management style is necessary to halt the continued decline of industry. We can no longer live with commonly accepted levels of delays, defective material and workmanship.
3   Cease dependence on inspection to achieve quality. Eliminate the need for mass inspection by building quality into the product in the first place using statistical techniques.
4   End the practice of awarding business on the basis of price tag. Purchasing must be considered with the design of product, manufacturing and sales to work with the chosen suppliers; the aim is to minimize total cost, not merely initial cost.
5   Improve constantly and for ever every activity in the company, to improve quality and productivity and thus constantly decrease costs.
6   Institute training and education on the job, including management.
7   Institute supervision. The aim of supervision should be to help people and machines to do a better job. Supervision must react to adverse conditions.
8   Drive out fear, through effective communication, so that everyone may work effectively for the company.
9   Break down barriers between departments. People in research, design, sales and production must work as a team to tackle anything that may be encountered with the product or service.
10  Eliminate slogans, exhortations and targets for the workforce asking for zero defects and new levels of productivity without provision of the methods to achieve this.
11  Eliminate work standards that prescribe numerical quotas for the day. Substitute aids and helpful supervision, using the methods to be described.
12a Remove the barriers that rob the hourly worker of the right to pride of workmanship. The responsiblity of the supervisor must be changed from sheer numbers to quality.

12b Remove the barriers that rob people in management and in engin-
    eering of their right to pride of workmanship. This means, *inter alia*,
    abolition of the annual or merit rating and of management by objec-
    tive.

13 Institute a vigorous programme of education and retraining. New skills
   are required for changes in techniques, materials and service.

14 Put everybody in the company to work, to accomplish the transformation. Define top management's commitment to improving quality and
   productivity.

In addition to the fourteen points, Deming has highlighted common obstacles
to achieving quality. These are the seven deadly diseases, of which the first
five are:

- lack of constancy of purpose
- emphasis on short-term profits
- evaluation of performance
- mobility of management
- running a company on visible figures alone.

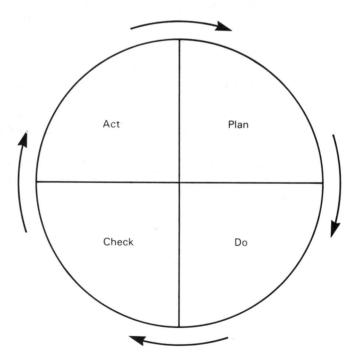

**Figure 20**  Deming cycle.

The other two relate to medical costs and warranty costs which relate primarily to the USA.

See also G is for Guru and H is for *Hoshin*.

# D is for Departmental purpose analysis

Departmental purpose analysis (DPA) is a technique, often attributed to IBM, that commences with looking at internal customer–supplier relationships, progresses into analysing value-adding and non-value-adding activities, and culminates with the setting of performance targets for internal customer–supplier interfaces.

See P is for Process management and performance measurement for an in-depth description of such an approach.

# D is for Design of experiments

Statistical design of experimental techniques to obtain maximum information from a small number of test runs originates from work performed in the 1920s by R.A. Fisher. More recently, however, this approach has been popularized by Genichi Taguchi through his work on designing to ensure quality.

### Taguchi

Dr Genichi Taguchi is executive director of the American Supplier Institute, Inc., and also director of the Japan Industrial Technology Institute. Born in 1924, he graduated from Kiryu Technical College, and received a doctorate in science from Kyushu University in 1962.

Taguchi joined the Electrical Communications Laboratory (ECL) of Nippon and Telegraph Company in 1949, and worked there until 1961 to improve the productivity of the ECL's research and development activities. He went to the USA in 1962 and visited Princeton University as a research associate. He returned to Japan and was a professor at Aoyama Gakuin University in Tokyo until 1982, during which time he served as a consultant to major Japanese corporations such as Toyota Motors, Fuji Film and Nippondenso.

His major contribution has involved combining engineering and statistical methods to achieve rapid improvements in costs and quality by optimizing product design and manufacturing processes.

Taguchi's techniques aim to go a stage further than the principles of eliminating causes of problems, by allowing the user to design products or processes that are insensitive to causes.

To achieve insensitive and optimized processes it is often necessary to study the effects of many product or process variables and their interaction with variables that may cause the process to deviate from its targets ('noises'). The study of all possible combinations of variables in an industrial process will, without the application of statistical techniques, involve many thousands of experiments and be extremely costly.

Statistically planned experiments use statistical tools such as orthogonal matrices to allow experimental designs to be produced that require far fewer experiments.

Whilst some statisticians may argue over the mathematical validity of some aspects of Taguchi's approaches, they are widely recognized and so are overviewed here in the context of examples. Space does not permit a complete detailed explanation, but readers will gain an appreciation of the possible applications of these powerful techniques.

## The evaluation of quality

Conventional methods of evaluating quality have often focused on how close to zero are returns of scrap, rework and its cost, warranty returns and costs and so on. The problem with this strategy is that it comes too late – the cost has been incurred.

A further measure of quality in current use is conformance to specification. Specifications are very important. They are a very efficient way for the designer to communicate his intent – the nominal dimension being the point at which the design functions optimally, and the upper and lower limits being the point at which the design function is unacceptable. Implicit in this is the understanding that deviation from design intent is a deviation from optimum performance.

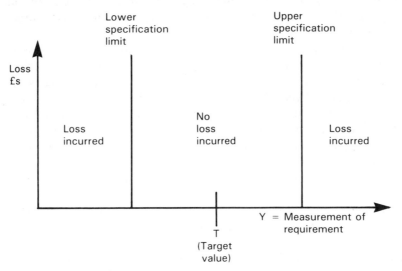

**Figure 21** Conventional thinking on acceptability.

Largely through the history of 'go and no-go' gauging, being 'in spec.' came to mean complete acceptability and to be synonymous with an evaluation of 'quality' within the manufacturing community. See Figure 21.

The introduction of processes with **capability** indices better than 1.33 and **statistical process control** enabled customers to perceive that products closer to design intent exhibited a better measure of 'quality'.

## Taguchi's loss function

In 1960 Taguchi had shown that if the design intent of the target value was the value at which the design, whether product or process, functioned best, then any deviation from that value meant an incremental deterioration of performance and incurred a loss. This loss was passed on to the next customer, whether internal or external, so that ultimately both the company and society would experience that loss in some way.

Close to the target value, the loss is minimal, but loss increases parabolically as the deviation increases to a point where the customer would find the product no longer acceptable and go for sourcing elsewhere. This concept is shown in Figure 22.

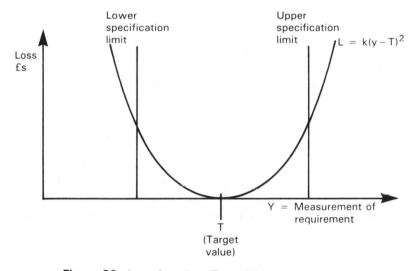

**Figure 22**  Loss function (Taguchi).

It can be seen that any deviation from target value of a critical characteristic may now be evaluated not only in dimensional terms but also in financial ones.

## Noise and variation

Variations from target values of critical characteristics affect the performance of any system and are a cost burden, as the loss function shows.

Variation is composed of two elements – first, the width of the spread around the average of the characteristic; and, second, the distance, if any, of the average from the target value. The impact of parameter design is to reduce these two elements towards zero.

Deviations either side of the average value – the spread – are usually caused by the largely uncontrollable effects on the system. For convenience these uncontrollable effects are called 'noise'.

Noise may be broken down into three categories:

- Outer noise – humidity, temperature, vibration, dust, layout etc.

- Inner noise – wear and deterioration.

- In-between noise – supplier material, component variation due to operator influences etc.

The strength of these noises largely determines the amount of spread, as the noises directly impact on those controllable parameters or factors that make the system happen – in a machining operation – feed, speed, cutter type and so on.

The reduction in spread – producing more parts nearer the average dimension – is more difficult to achieve than moving the average to the target value, which can usually be done by correcting the setting of one or more parameters on line; for example, increasing or decreasing speed, temperature and so on.

The reduction in spread is more suitably an off-line activity, achieved by a defined system of experimentation to identify those controllable parameters that are significant in controlling variation and to identify the best setting of each parameter so that the system is robust to the noises affecting the system.

Taguchi's method of testing controllable factors against the effects of noise factors achieves this property of robustness. Robustness ensures that controllable factors do produce identified critical characteristics with minimum deviation day after day, month after month. This results in a consistent performance of the system in the hands of the consumer.


## Experimental design and orthogonal arrays

Experimental design involves testing different combinations of the various levels of controlling factors in a defined small-scale logical sequence, under the noise conditions that, for the purposes of the experiment only, need to be controlled.

Table 3. Experimental results from an injector moulding development, using an L8 orthogonal array

| Experiment no. | Controllable factors | | | | | | | Noise factor | | | | Average | S/N |
|---|---|---|---|---|---|---|---|---|---|---|---|---|---|
| | Material type A | Mould temp. B | Clamp time C | Screw speed D | Injection pressure E | Injection speed F | Back pressure G | M/C1 night-shift | | M/C2 day-shift | | | |
| 1 | 1 | High | 6 | Low | 1200 | Slow | 60 | 1.501 | 1.500 | 1.494 | 1.495 | 1.497 | 52.596 |
| 2 | 1 | High | 6 | High | 1400 | Fast | 100 | 1.503 | 1.502 | 1.498 | 1.501 | 1.501 | 56.838 |
| 3 | 1 | Low | 3 | Low | 1200 | Fast | 100 | 1.499 | 1.498 | 1.496 | 1.495 | 1.497 | 58.276 |
| 4 | 1 | Low | 3 | High | 1400 | Slow | 60 | 1.503 | 1.501 | 1.501 | 1.500 | 1.501 | 61.533 |
| 5 | 2 | High | 3 | Low | 1400 | Slow | 100 | 1.506 | 1.504 | 1.500 | 1.499 | 1.502 | 53.154 |
| 6 | 2 | High | 3 | High | 1200 | Fast | 60 | 1.504 | 1.502 | 1.498 | 1.497 | 1.500 | 53.142 |
| 7 | 2 | Low | 6 | Low | 1400 | Fast | 60 | 1.506 | 1.505 | 1.499 | 1.500 | 1.502 | 52.625 |
| 8 | 2 | Low | 6 | High | 1200 | Slow | 100 | 1.505 | 1.503 | 1.500 | 1.502 | 1.502 | 57.169 |

From the results of these tests, a choice of factor level setting for each factor may be made, which will give the near optimum critical quality characteristic being approached as well as having the result that the system will be insensitive to the effect of the noises studied, variation around that optimum value being minimized.

Convenient testing methods are the orthogonal arrays, originally developed by Sir Roland Fisher in the 1920s. The orthogonal array is a very efficient method of achieving a near optimum result in a relatively small-scale experiment. Typically an eight-run test can reproduce practically the same useful information as a 128-combination test. Table 3 shows the results of a typical experimental layout using an L8 orthogonal array. This example is for an injection moulding development and shows how information is gained to assist optimization of the process being investigated.

The preferred steps in the experimental phase are as follows:

1  Define the problem.
2  Determine the objectives.
3  Identify critical response characteristics which are measurable.
4  **Brainstorm** to identify all factors affecting performance.
5  Separate the factors into control and noise factors.
6  Decide how noise factors can be measured for experimental purposes.
7  Determine control factor levels and values.
8  Design the experiment:
   ● select the orthogonal array
   ● assign factors and interactions (if any) to columns
   ● select an outer array for noise factors.
9  Conduct the experiment and collect the data.
10  Analyse the responses for signal/noise and averages.
11  Interpret the results of the analysis.
12  Select optimum conditions of control factors.
13  Predict the outcome of the recommended choice of levels.
14  Run a confirmation experiment to verify the prediction.

Readers wishing to review further the benefits achievable through such approaches will require more detail than this conceptual overview and should consult the bibliography for suitable material.

# E is for Employee participation

## Employee participation and involvement

This section starts by discussing the need for employee participation and progresses to discussing typical structures of reporting that will assist you in achieving it. Successful Total Quality Management programmes have repeatedly shown that employees 'buy into' the changes needed, understand and achieve objectives and work towards demanding standards if they have been involved in understanding the need and identifying the method to satisfy those needs.

Enthusiastic employee involvement cannot be achieved by the adoption of one or two 'magic' tools and techniques. Such participation is achieved from the top downwards and often involves significant attitude changes for the managers involved. These changes, in many organizations, will mean a shift from the Tayloristic, or scientific, school of management, based on highly simplified jobs with task repetition, the outcome being an organization where devolution of responsibility and the achievement of staff, acting under their own empowerment, occurs at the lowest possible levels.

The Industrial Society has quantified their opinion of the elements of successful employee involvement. These are listed below and are often pre-requisites for success for any Total Quality Management programme:

- Briefing all employees on a regular basis about matters at work, including the needs, plans and achievements of both the organization and the local work unit.
- Providing appropriate financial and performance indicators.
- Consulting employees before decisions are taken on matters which affect them, feeding back their views and ideas to the decision takers and then explaining the reasons for the final decision.
- Delegating the organization of work, and of the working environment, to as low a level as is consistent with other objectives, such as the management of quality.
- Recognizing employees' contribution to the performance of the organization by providing them with a financial share in its success.

When you consider the above elements, certain themes may be picked out. These themes form the basis of good, effective employee participation and are as follows:

1 communication
2 organization and structure
3 recognition and reward
4 teamwork and teambuilding.

Each of these items is a subject in itself; refer to each elsewhere in the Executive Encyclopaedia for a more detailed account (C is for Communication, O is for Organization and structure, R is for Recognition and reward and T is for Teamwork and teambuilding.

# F is for Failure mode and effects analysis

## Potential failure mode and effects analysis (FMEA)

Failure mode and effects analysis is a simple method of reviewing product designs or processes to discover any inherent weaknesses and eliminating them. It involves the identification of all possible failure modes, their effects and the causes of these failures. Typically such techniques have been applied to product design and manufacturing processes. However they are increasingly being applied in non-manufacturing areas.

| | |
|---|---|
| Failure mode | how the component fails to fulfil its function. |
| Failure effect | what happens to the system and what the customer experiences when a failure occurs. |
| Cause of failure | the reason why the failure mode occurred. |

Failures are rated on the basis of probability of occurrence, severity and probability of detection to give an improvement weighting which indicates priorities for improving the product.

The FMEA is now a mandatory part of many contracts, particularly in automotive sectors, because of its effectiveness in the face of the changed legal and market environments.

## Uses of an FMEA

1  It identifies design or process weaknesses and allows for preventive actions to occur.
2  It sets priorities for action.
3  It defines responsibility for action.
4  It stops design or process weaknesses being passed on to the next stage if an FMEA is done at each stage; for instance, concept, intermediate and final stages.
5  FMEAs are typically done on all safety critical components.
6  FMEAs should also be done on:

- all new components
- components carried over from previous designs but in a new environment
- modified components.

7  It can form the basis of:
- the quality plan
- the design record
- the maintenance plan
- diagnostic routines
- the maintenance manual.

8  It can be used as the basis for some other design review techniques such as:
- fault trees

- reliability block diagrams
- simulation.

9 By using quantitative ratings, improvement during design and development can be monitored.

10 The effects of design changes and modifications can be assessed.

11 An FMEA for a previous component can provide a start for an FMEA on a similar new component.

12 It identifies the critical characteristics of the design which require controls such as **statistical process control**.

FMECA is a refinement which introduces 'criticality' by rating the probability and seriousness of failure so that a numerical ranking of failures is possible. Figure 23 is an example of documentation for use in FMEA.

See also P is for Problem solving.

## F is for Force field analysis

Force field analysis is a method to focus your staff's thought processes on the factors that can affect an activity or proposed change positively or negatively. The relative strength of these forces can be established by ranking on a numerical scale, and are typically shown graphically, on a chart such as Figure 24.

## G is for Guru

### Quality guru

A dictionary definition of a guru is 'an authority, a respected instructor or a religious teacher'. Quality gurus are all of these, but they may also wish to be known as prophets, each with their own set of commandments. This part has looked at, in detail, the contribution of the three most widely recognized gurus – P.B. **Crosby**, W.E. **Deming** and J.M. **Juran**.

There are many others who have been awarded similar status to these. Your attention is drawn to the five names below and to the references in the bibliography.

#### Armand V. Feigenbaum

Feigenbaum was one of the pioneers in quality improvement. In 1961 he produced the reference book, *Total Quality Control: Engineering and Management.*

#### Professor Kaoru Ishikawa

Ishikawa developed a number of tools used in quality improvement schemes. These include the concept of quality circles and the **cause and effect diagram**.

#### Tom Peters

Peters has studied and reported on quality in a number of companies and synthesized contributions to excellence. He has identified the two ways of sustaining superior behaviour as:

| Part or process | Potential failure mode | Likely effect of failure | Cause of failure | Probability factor A | Severity factor B | Detection factor C | Improvement weighting A x B x C |
|---|---|---|---|---|---|---|---|
|  |  |  |  |  |  |  |  |
|  |  |  |  |  |  |  |  |
|  |  |  |  |  |  |  |  |
|  |  |  |  |  |  |  |  |
|  |  |  |  |  |  |  |  |
|  |  |  |  |  |  |  |  |
|  |  |  |  |  |  |  |  |
|  |  |  |  |  |  |  |  |
|  |  |  |  |  |  |  |  |
|  |  |  |  |  |  |  |  |
|  |  |  |  |  |  |  |  |
|  |  |  |  |  |  |  |  |
|  |  |  |  |  |  |  |  |

**Figure 23**  Failure mode and effects analysis.

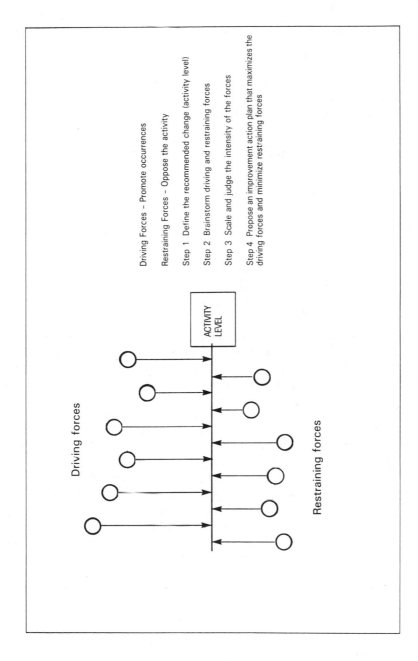

**Figure 24** Force field analysis.

1 taking care of your customers via superior service and quality
2 constant innovation.

Peters developed the concept of management by wandering around (MBWA).

### Dr Genichi Taguchi

Taguchi is the Japanese founder and executive director of the American Supplier Institute and the holder of several Deming prizes. His major contributions are the principles of the quality loss function and the **design of experiments.**

# H is for *Hoshin*

*Hoshin Kanri* is Japanese and its literal business translation means 'policy deployment'. It is a method for deploying annual strategic plans down through an organization.

*Hoshin* Plans are sub-divided into two major parts:

- breakthrough
- business fundamentals.

Let us look at each in turn, as shown in Figure 25.

### Breakthrough

The chief executive and the management team should review the company's measurement indicators in order to identify the key issues affecting the business for that year. Inputs to this should be sought from long-term plans, the economic state of the business, customer and employee satisfaction matters, and a review of last year's achievements.

Once the review has been done, consensus on the key business issue is shared within the team. It is then that the chief executive and the management team will agree the 'breakthrough' objective for the company for the next year. Up to three objectives may be set. From the objectives a maximum of five or six key strategies will be agreed and owners assigned. The owners of these strategies will view these as their objectives, and will develop strategies and owners for each. Again, these strategies will then become the next level's objectives.

You can see from this cascade method that the policy (i.e. objective) is cascaded (deployed) through the organization. There is a danger that the whole plan could be 'deployed' leaving everyone at the top of the organization with nothing to do! Therefore it is important that *Hoshin* Plans are deployed and not just dropped through to the next level down. The chief executive and his team should own at least one strategy themselves and so on.

A critical part of any planning process is the review. 'Did we achieve what we set out to do? *Hoshin* uses a regular – usually quarterly – method of reviewing plans. Plans are reviewed bottom up so, by the time they reach the top, the manager accountable for the plan should know whether the annual objective is achievable. With this in mind, it is very important that the correct measures

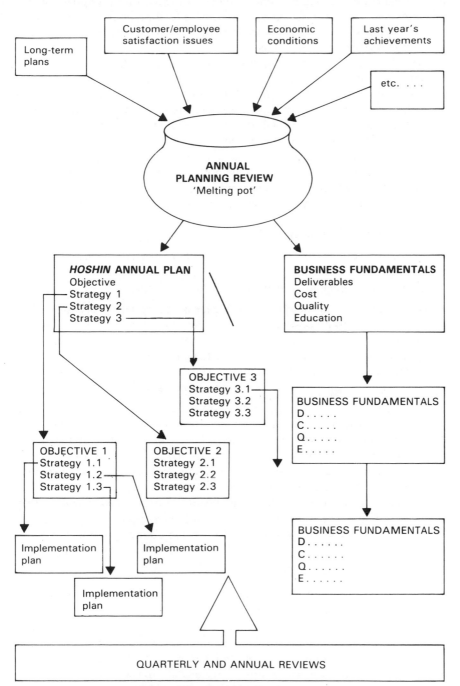

**Figure 25** *Hoshin* planning process.

are assigned to the strategies right from the start. Measures (not goals) should be achievable, easy to collect and to chart. Also, there is no point in measuring for measuring's sake. Decide what the key measure for the strategy will be and report that on a regular basis.

Once the *Hoshin* Plan is deployed to about the third tier, the plan changes from objectives and strategies to strategies and tactics. Strategies say what you want to achieve, tactics say how and by when. At this level, you would use an 'implementation plan' which shows the strategy, tactics and time scales. It is important that the owner of the tactic knows the deadline and can manage the task into his daily diary.

Finally, one of the most important considerations with any plan is to ensure that it doesn't become the annual business 'trophy' that is pinned up on the wall for the first month and then filed in the bottom drawer! Plans should be well deployed, launched, communicated and reviewed. The chief executive must ensure that this happens from the top-level review, which must not be delegated.

## Business fundamentals

These are the 'business as usual' measures that you need to make and monitor as part of everyday business life.

Business fundamentals are invariably split into four.

- Deliverables – The output of the business/department/process
- Costs – Includes all cost-related items you as a manager have to manage, like manufacturing costs, productivity, selling costs, people costs etc.
- Quality – which includes customer satisfaction, product/process quality etc.
- Education – Which is all the human resource measures that are important like training, employee development, appraisal and satisfaction.

By looking at these four measures you are ensuring that you are looking at all of the dimensions of the business and are unlikely to miss out anything that is important. Once you have decided what the business fundamentals are, then you should monitor and chart them regularly. By doing this you will be able to predict issues and successes based on real data and take corrective action before problems become too serious.

Business fundamentals should be the 'health' measure of the business, they should flag issues and subsequent process improvement activities. They are essential for business success and will keep the company competitive.

# J is for Juran

Dr Joseph M. Juran is renowned as a world-class expert and for his contribution to quality improvement.

Juran set out as an engineer working for American Telephone and Telegraph, who were early pioneers of statistical methods of quality. Juran worked his way up to corporate industrial engineer with Western Electric before branching out on his own. In 1951 he published the *Quality Control Handbook*, which has since become a standard reference work on the subject. Like Deming,

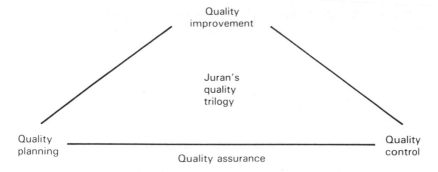

**Figure 26**   Juran's quality trilogy.

Juran was invited to conduct a lecture tour of Japan. This took place in 1954. He was, also like Deming, awarded the Second Order of the Sacred Treasure by the Emperor of Japan. Juran is chairman of the Juran Institute, which offers training and consultancy in quality management.

Juran has developed his view of quality from a statistical base to a management philosophy to be integrated with other management strategies. He has drawn parallels between quality processes and financial processes in the control of a business. These quality processes are summed up by the quality trilogy: quality planning, quality control and quality improvement. These are illustrated in Figure 26.

Looking at these basic quality processes, the first is quality planning – preparing to meet quality goals. The key points are:

- Identify the customers, both external and internal.
- Determine customer needs.
- Develop product features that respond to customer needs (products include both goods and services).
- Establish quality goals that meet the needs of customers and suppliers alike, and do so at a minimum combined cost.
- Develop a process that can produce the product features needed.
- Prove process capability – prove that the process can meet the quality goals under operating conditions.

The second basic process is quality control, which is the means of meeting the quality objectives. This revolves around the following actions:

- Choose control subjects – what to control.
- Choose units of measurement.
- Establish measurement.
- Establish standards of performance.
- Measure actual performance.
- Interpret the difference (actual versus standard).
- Take action on the difference.

The final process is that of improvement – breaking through to new levels of performance. The methodologies for this process are:

- Prove the need for improvement.
- Identify specific projects for improvement.
- Organize to guide the project.
- Organize for diagnosis – discovery of causes.
- Diagnose to find the causes.
- Provide remedies.
- Prove that the remedies are effective under operating conditions.
- Provide for control to hold the gains.

See also G is for Guru.

## K is for *Kaizen*

*Kaizen* is a Japanese word meaning 'slow, never-ending improvement in all aspects of life'. It represents a Japanese approach to improvement and can be interpreted as continuous improvement by everyone in all areas. Typically *kaizen* has been at the heart of quality improvement in Japanese companies, but it has also, because of what it is, been instrumental in all areas of improvement including productivity, stock reduction and maintenance.

*Kaizen* differs from the classical Western approach to improvement principally in that it relies on an investment in people, not equipment or systems. Innovation has been the classical Western approach to improvement; large sums of money have been spent on new equipment and systems using the latest technology and requiring specialist involvement to give large-step changes in performance. Undoubtedly this has led to dramatic improvements but typically they have not been standardized and maintained. This has resulted in a fall in performance over a period of time. This can be illustrated as shown in Figure 27.

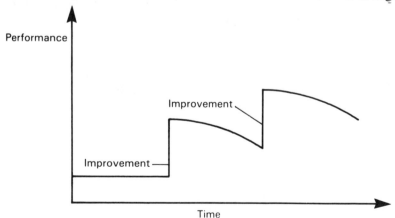

**Figure 27**   Improvement through innovation steps may not last.

*Kaizen*, on the other hand, is a continuous series of small-step improvements made on existing equipment or systems by the people who actually work in that area. It is based on existing techniques and technology without the need for spending relatively large sums of money. It does not rely upon specialist involvement but, if required, it will be used to support those directly involved in making the improvement. An important aspect of *kaizen* is the standardization and maintenance of improvements, this is as crucial to the process as the improvement itself. Improvements must become standardized and maintained until further improvement is made. Overall this pattern of small-step improvements followed by standardization and maintenance, which on their own are insignificant, gives an improvement profile as shown in Figure 28.

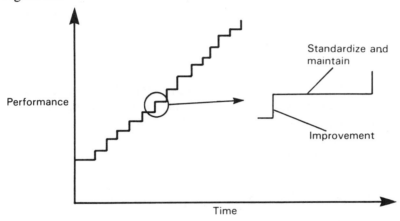

**Figure 28**   Continuous improvement through *kaizen*.

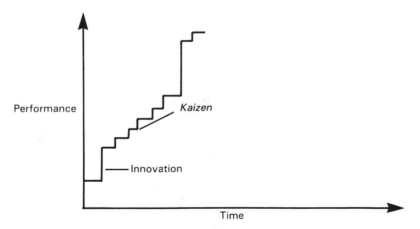

**Figure 29**   Innovation supported by *kaizen*.

*Kaizen* is most effective when used in combination with innovative-type improvements. Even the continuous nature of *kaizen* is eventually not sufficient on its own; there may come a time when an innovative step is required, but this must be further supported by a *kaizen* approach. This will give the combination shown in Figure 29.

There is a structured approach to *kaizen* improvements. It is important to follow each and every step, in particular the final step, to ensure that the improvement is a lasting one, or at least continues until there is further improvement:

1  Define area for improvement.
2  Analyse and select appropriate problem.
3  Identify causes.
4  Plan countermeasures.
5  Implementation.
6  Confirmation of result.
7  Standardization.

Compare this with the classical problem-solving approach and you will see that it is very similar, except that the *kaizen* approach stresses the importance of standardizing improvments.

It is important to note that *kaizen* is applicable to all levels in an organization, both on an individual or team basis.

# M is for Management systems

Many companies see quality management systems and third-party certification as one step along the road to Total Quality. Others perceive it as Total Quality in itself. There are many benefits in using quality systems; the following list highlights just a few of the possibilities:

- Offer assurance.
- System not people oriented.
- Establish fundamental controls.
- Freeze reversible changes.
- Develop interfaces with customers/suppliers.
- Establish monitoring through internal audits.
- Introduce the corrective action concept.

However, there are also dangers. If wrongly applied, quality management systems can become very bureaucratic and can be seen as a restraining force against ongoing and innovative improvement.

The following history of quality systems, leading to **BS5750**, is a starting-point for those considering integrating a systems approach into their initiative. Further sources of help are contained in the bibliography.

# History of quality management systems leading to BS5750/ISO9000/EN29000

After the Second World War pressure for quality came from the military and resulted in the 05 series MOD quality standards and the AQAP series of NATO standards. Major companies, notably in the automotive industry, began publishing their own quality system standards and assessing their suppliers in line with them. In an attempt to avoid the proliferation of quality system standards and to reduce multiple assessments, BSI developed the military standards into a series which could be used in companies supplying commercial markets. This eventually resulted in the publication of BS5750: Parts 1, 2 and 3 in 1979.

Pressure for an international quality system standard resulted in the publication in 1987 of the series of ISO9000 standards, which were established, using BS5750, as a base for discussion by the international standards committees. After publication of this standard, the British Standards Institution (BSI) revised BS5750: Parts 1, 2 and 3 so that they were identical to the ISO9000 standard and reissued it in June 1987. The new standard has been used for assessments

*Table 4.* International compatibility of the ISO9000 series

| Standards body (country) | Quality management and quality assurance standards. Guidelines for selection and use. | Quality systems. Model for quality assurance in design/development, production, installation and servicing. | Quality systems. Model for quality assurance in production and installation. | Quality systems. Model for quality assurance in final inspection and test. | Quality management and quality system elements. Guidelines. |
|---|---|---|---|---|---|
| ISO | ISO9000: 1987 | ISO9001: 1987 | ISO9002: 1987 | ISO9003: 1987 | ISO9004: 1987 |
| Australia | AS3900 | AS3901 | AS3902 | AS3903 | AS3904 |
| Austria | ö Norm ISO9000 | ö Norm ISO9001 | ö Norm ISO9002 | ö Norm ISO9003 | ö Norm ISO9004 |
| Belgium | NBN X 50-002-1 | NBN X 50-003 | NBN X 50-004 | NBN X 50-005 | NBN X 50-002-2 |
| Canada | CSA Z2990-85 | CSA Z2991-85 | CSA Z2992-85 | CSA Z2994-85 | CSA Q420-87 |
| Denmark | DS/ISO9000 DS/EN29000 | DS/ISO9001 DS/EN29001 | DS/ISO9002 DS/EN29002 | DS/ISO9003 DS/EN29003 | DS/ISO9004 DS/EN29004 |
| Finland | SFS-ISO9000 | SFS-ISO9001 | SFS-ISO9002 | SFS-ISO9003 | SFS-ISO9004 |
| France | NFX 50-121 | NFX 50-131 | NFX 50-132 | NFX 50-133 | NFX 50-122 |
| Germany (FR) | DIN ISO9000 | DIN ISO9001 | DIN ISO9002 | DIN ISO9003 | DIN ISO9004 |
| India | IS10201 Part 2 | IS10201 Part 4 | IS10201 Part 5 | IS10201 Part 6 | IS10201 Part 3 |
| Ireland | IS300 Part O/ ISO9000 | IS300 Part I/ ISO9001 | IS300 Part 2/ ISO9002 | IS300 Part 3/ ISO9003 | IS300 Part O/ ISO9004 |
| Netherlands | NEN ISO9000 | NEN ISO9001 | NEN ISO9002 | NEN ISO9003 | — |
| Norway | — | NS5801 | NS5802 | NS5803 | — |
| South Africa | SABS0157: Part 0 | SABS0157: Part 1 | SABS0157: Part 2 | SABS0157: Part 3 | SABS0157: Part 4 |
| Spain | UNE 66900 | UNE 66901 | UNE 66902 | UNE 66903 | UNE 66904 |
| Switzerland | SN ISO9000 | SN ISO9001 | SN ISO9002 | SN ISO9003 | SN ISO9004 |
| United Kingdom | BS5750: 1987 Part 0: Sect 0.1 ISO9000/EN29000 | BS5750: 1987 Part 1 ISO9001/EN29001 | BS5750: 1987 Part 2 ISO9002/EN29002 | BS5750: 1987 Part 3 ISO9003/EN29003 | BS5750: 1987 Part 0: Sect 0.2 ISO9004/EN29004 |
| USA | ANSI/ASQC Q90 | ANSI/ASQC Q91 | ANSI/ASQC Q92 | ANSI/ASQC Q93 | ANSI/ASQC Q94 |
| Yugoslavia | JUS AK 1.010 | JUS AK 1.012 | JUS AK 1.013 | JUS AK 1.014 | JUS AK 1.011 |
| European Community | EN29000 | EN29001 | EN29002 | EN29003 | EN29004 |

carried out since 1 October 1987. Subsequently, the European standard EN29000, which is identical and equivalent to ISO9000, has been issued.

The international compatibility of the ISO9000 series is shown in Table 4. The 1993 ISO Document 'Vision 2000' states the belief that ISO9000 standards should continue to be the basis of quality standardization into the twenty-first century.

## Part of the standard – the first decision

BS5750: Part 0, a new addition to the standard, has also been issued in two sections, again to be identical to parts of the ISO series. This part is issued only as a guide and not for use in assessing the basis of a contract.

BS5750: Part 0: Section 0.1: 1987  – Guide to selection and use
ISO9000 – 1987
EN29000

BS5750: Part 0: Section 0.2: 1987  – Guide to quality manage-
ISO9004                                                ment and quality system
EN29004                                              elements

BS5750: Part 1: 1987                       – Specification for design/
ISO9001 – 1987                               development, production,
EN29001                                          installation and servicing

BS5750: Part 2: 1987                       – Specification for produc-
ISO9002 – 1987                               tion and installation
EN29002

BS5750: Part 3: 1987                       – Specification for final
ISO9003 – 1987                               inspection and test
EN29003

An organization is certified to Parts 1, 2 or 3, according to what is most appropriate to its business.

## Other decisions to be taken

When deciding on the type of system to implement, as well as which part of the standard to apply, there are some other points to consider.

### Scope of registration

You must decide which products and/or services you wish to have covered by your registration. There may be some products at the lower end of your market for which your customers' requests are more for flexibility in quick turn-round and less on meeting a system standard. An example is the customer who gives you a 'fag packet' drawing and says 'make me half a dozen of those by lunchtime'. The interests of your company and your customer may not be best served by including these products in your system that is to be assessed. You must agree with the assessing body your scope of registration before the assessment.

### Sector or individual firm scheme

You should consider whether to apply under an industry sector-based scheme or an individual firm certificate scheme. When assessment and registration to BS5750 was first introduced, a number of quality assessment schedules (sometimes referred to as technical schedules) were drawn up. Each related to a particular industry sector and were drawn up as a secondary set of requirements which helped to relate the general BS5750 standard to that particular industry.

With the launch of the National Quality Campaign by the government in 1983 and the financial assistance schemes operated by the Department of Trade and Industry, the demand for registration increased dramatically with many individual companies which were not covered by the scope of existing quality assessment schedules coming forward for registration. To cope with the increase in demand, individual firm assessment schemes were established. Under these schemes, no quality assessment schedule is required and assessment is carried out with reference only to the relevant part of the standard and to the company's quality manual. Unless your major customers have a preference for a sector scheme, an individual firm scheme will be easier to implement.

### Third-party assessors

You should then go on to consider which third-party assessment body to apply to for registration. Although BS5750 is naturally associated with the BSI, there are several other bodies who can carry out assessment to BS5750, notably Lloyd's Register and Yarsley. Again, your customers' preference should be sought before making a decision. You should also note whether a third-party assessor is a recognized assessment body for the scope to be assessed. In 1985 the government established the National Accreditation Council for Certification Bodies (NACCB) which assesses the assessors. It is recommended that you choose a body accredited by the NACCB, but also one which is recognized by the industry either you or your customers belong to.

### NACCB-approved bodies (the first eight)

001 – Lloyd's Register Quality Assurance (LRQA)
002 – UK Certification Authority for Reinforcing Steel (CARES)
003 – BSI Quality Assurance
004 – British Approvals Service for Electrical Cables (BASEC)
005 – Yarsley Quality Assured Firms (YQAF)
006 – Ceramic Industries Certification Scheme
007 – Loss Prevention Certification Board
008 – Bureau Veritas Quality International (BVQI)

Contact details for the NACCB are given in the bibliography.

# O is for Organization and structure

## A typical Total Quality Management structure

Total Quality will not happen unless a structural framework is established which allows everyone to become involved and which facilitates good communication relating to quality improvement. (This does not mean radically revising the current management structure or undermining the normal line management authority!) It should be integrated with the normal management structure, providing a means by which management and supervision can gradually be freed from the role of 'firefighter' that many hold and move more towards that of a planner, trainer and coach. See Figure 30.

To lead a successful Total Quality initiative, as we highlighted before, there is a need for senior management to demonstrate a highly visible commitment to quality. This will not be achieved simply by putting an item relating to quality on the agenda for board or senior management meetings. There is a need to establish a senior management quality steering group which is dedicated to the implementation, maintenance and monitoring of the Total Quality programme. Devoting the necessary time and energy to membership of this group goes part of the way to demonstrating commitment by senior management.

In smaller companies, the steering group will consist of the entire board of directors or senior management team. In larger companies, the steering group may report to the board of directors. However, it is important that the

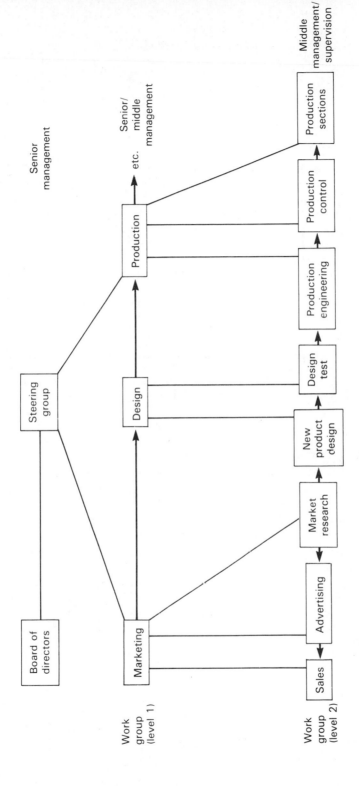

**Figure 30** Relationship between functional structure and TQM work groups.

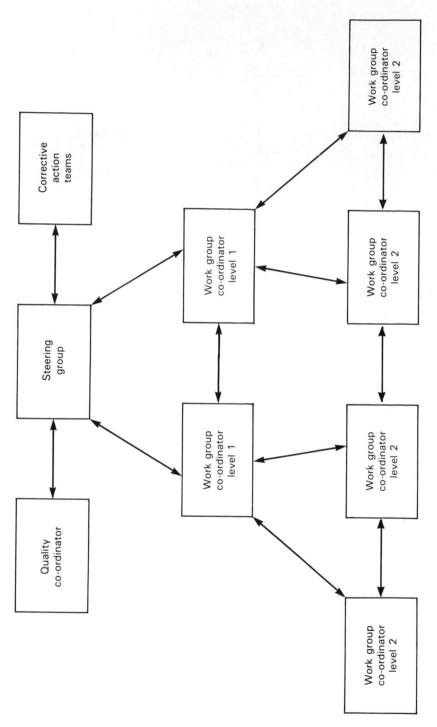

**Figure 31** Structure for co-ordination of TQM work groups.

steering group is led by the chief executive and includes members from a senior level within the organization. It is also important to ensure that the need to delegate responsibility for quality down throughout the organization does not result in the abdication of responsibility by senior managers.

Before we go too far in defining our structure for managing quality at all levels in the company, we must address a burning issue which arises in many companies. What is the role of the quality manager and his/her department in the company's changing approach to quality? Depending on how the current role of the function is perceived within the company, you may need to change the emphasis from one of criticism to one that is more in support of other functions. There should be a gradual change from predominantly inspection activities to those of auditing and quality engineering. For example, goods receiving inspectors may become supplier quality assurance engineers.

While some inspection and testing may still be required because of the type of industry, product or processes, this need not be carried out by a centralized quality function. It can be more effective to have the inspector as part of the production team, supplying them with feedback rather than offering external criticism. In many companies operator self-certification has been adopted with, in some cases, inspectors being redeployed as production operatives.

There is no one structure which can be laid down for all organizations, but below is one approach to fostering participation, improved communication and working for internal customer satisfaction.

See also E is for Employee participation, C is for Communication and T is for Teamworking and teambuilding.

There is no single model that will fit every organization's need to have a structure that maximizes communication and involvement. However, there are typical approaches. One such approach is shown in Figure 31. In this representation of a Total Quality Management structure, the following roles and responsibilities are established.

### Steering group

As the driving force for the initiative, its role is to ensure the successful implementation of the policies and objectives set for the company. Included in its responsibilities are to:

- agree objectives with the level-1 work group co-ordinators
- review performance of work groups
- communicate the company's quality performance and the plans to improve it down to level-1 work groups
- respond to feedback on barriers to corrective action

- set priorities for corrective action projects
- select corrective action teams and agree objectives for them
- monitor the progress of corrective action teams.

### Quality co-ordinator

This is normally a full-time role, often taken by the quality manager. As well as being a member of the steering group, his/her duties may include:

- acting as internal consultant to the steering group
- acting as facilitator for group activity
- researching externally on quality information and techniques
- planning and co-ordinating quality training
- co-ordinating management auditing activity
- analysing and reporting on quality performance data.

### Work group co-ordinator (level 1)

This role is normally taken by members of senior or middle management who have responsibility for a particular function. They form a work group with members of management and supervision who report to them. They are responsible for:

- setting objectives for their work groups
- making resources available for quality improvement
- determining and meeting training needs
- communicating upwards on work group performance and downwards on company performance and plans
- resolving barriers to corrective action with other level-1 work group co-ordinators.

### Work group co-ordinators (level 2)

This role is normally taken by middle managers or supervisors. In some companies the work groups are encouraged to choose a co-ordinator not in a management/supervisory position. In these latter cases, the work group co-ordinator becomes a very useful support for communications, both internal and external, to the work groups. The co-ordinators' responsibilities will include:

- identifying internal customer work groups

- agreeing internal customer requirements
- establishing performance measures with their work group
- communicating with internal supplier work groups on corrective action
- communicating upwards on work group performance and downwards on company performance and plans.

### Corrective action teams

Work groups, by monitoring their performance measures, can institute corrective action at their level to improve their performance continually and thus increase internal customer satisfaction. Through their co-ordinators, problems created by their internal supplier work groups can also be resolved. Where additional resources or authority for corrective action between different functions are required, this can be resolved by level-1 work group activity.

When barriers to quality improvement extend outside the work group or indeed beyond the interface of two functions, these problems are communicated up to the steering group. To avoid stretching the resources of the company, the steering group will set priorities for solving these problems. They will then select a corrective action team to match the nature of the problem. The steering group will set the objectives for the corrective action team and monitor their progress.

Corrective action teams are:

- appointed by the steering group
- multifunctional
- tasked with solving specific problems
- required to report back to the steering group
- disbanded after the problem is solved.

## P is for Pareto analysis

Pareto's law was developed at the end of the nineteenth century when Vilfredo Pareto, while studying the concentration of wealth and income in his native country, Italy, found that a very large percentage of the total national income was concentrated in the hands of about 20% of the population. Pareto had worked for twenty years as an engineer and was mathematically oriented, so his natural inclination led him to express this concentration of income in mathematical terms. Cumulative curves were developed from the income–number of persons curves. These cumulative curves will be the basis of the following discussion.

For many years this relationship was considered to be an interesting phenomenon with very little practical use. However, shortly before the

Second World War, inventory control analysis revealed that, when inventory items were plotted on cumulative percentage graphs in order of descending value, Pareto's relationship seemed to emerge. It was observed that 10–20% of the items in a given inventory accounted for 80–90% of the total value of the inventory. The remaining large number of items then accounted for a very small portion of the inventory value.

Observations in many areas have shown a widespread applicability of Pareto's law. Examples are:

- only a few children cause problems in school
- only a few people account for most absenteeism
- only a few defects account for most quality losses.

The Pareto diagram is a graphical representation of the law. The various categories are listed across the bottom of a graph, then the cumulative totals are plotted as percentages, starting with the largest number to the left. In this way a diagram as shown in Figure 32 is formed. An alternative means of displaying the same information is in histogram form.

It is clearly seen that a small portion of activities are more important and

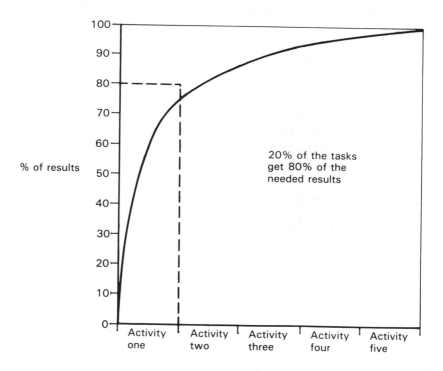

**Figure 32** The Pareto principle ('80/20' rule).

contribute most towards the objective. A large proportion are trivial in their contribution.

Pareto analysis enables prioritization of problems or areas for improvement. If resources are scarce, it is important that they are directed to where most benefits can be gained. The example in Figure 33 shows a practical application in histogram form.

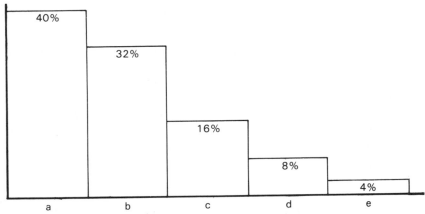

a = parts assembled incorrectly
b = wrong parts fitted
c = parts missed out
d = faulty parts fitted
e = damaged parts

**Figure 33**  Prioritizing problems using the Pareto principle.

Depending upon the objective, the percentages shown may refer to the number of interfaces, the cost of failure or time lost. It is important to recognize the objective behind the analysis and use the appropriate measurement. For example, if reducing quality costs was the objective in the situation above, the percentages would have to refer to costs of failure if they were to be useful.

See also P is for Problem solving and S is for Seven statistical tools.

## P is for Poka-yoke

One of the prime objectives of anyone carrying out an activity is to deliver defect-free products or services. Traditionally, inspection departments were used to ensure quality; however, even this end-of-the-line method cannot ensure a 100% quality product or service.

Poka-yoke devices are designed to back up the mistakes in operation that are made, which are inevitable owing to human nature. They prevent defects at source, ensuring the problem is either not allowed to happen or, if it does, stopping the operation to ensure it is recognized before it passes down the line with the possibility of greater losses.

In order to gain maximum benefit from poka-yoke devices, they should be developed as part of a continuous improvement approach by everyone from designers to production staff, and ideas should be shared by everyone, particularly those doing similar tasks.

Poka-yoke devices provide a low-cost method, ensuring 100% inspection at source. Typical examples are shown below:

1 Attach a fixture to a machine to prevent a workpiece being fitted the wrong way round.
2 Use limit switches to monitor procedures. If not performed correctly, operation is stopped.
3 Colour code similar parts; for example, right is blue, left is yellow.
4 Use compartments when counting. If an empty compartment appears before the end of the operation or a full compartment is left at the end, then the count must be incorrect.

# P is for Problem solving

Many organizations benefit from having a clearly defined methodology for problem solving. There are many individual approaches to problem solving but Figure 34 identifies seven key steps and the likely inputs that are present in all good problem-solving methodologies.

Several of the most common techniques are dealt with elsewhere in the Executive Encyclopaedia; see also A is for Auditing, F is for Failure mode and effects analysis, B is for Brainstorming, C is for Cause and effect diagram S is for Seven statistical tools, P is for Pareto analysis and Q is for Quality costs.

# P is for Process flow analysis

As part of an analysis of a process to identify opportunities for improvement, it is not uncommon to find that the 'workforce' staff have been involved in an analysis of their work processes by methods such as flow charting or process flow analysis.

Flow charting is often the first step. The commonly used symbols are sketched in Figure 35.

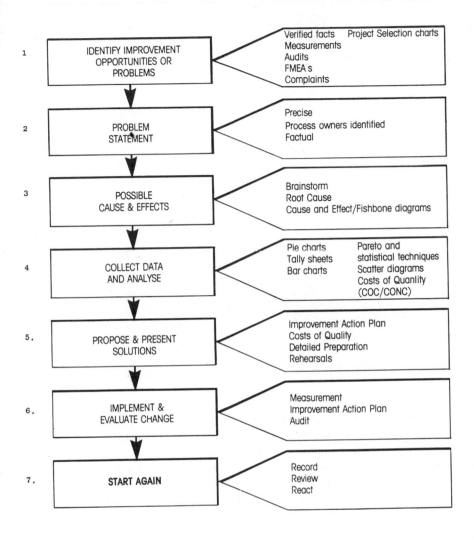

**Figure 34** An overview of problem-solving steps and likely inputs.

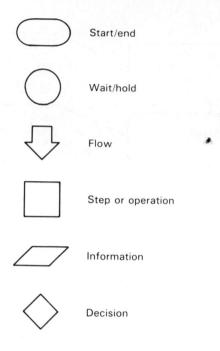

**Figure 35**  Symbols commonly used in flow charting.

Complex methods, such as IDef.O, exist for mapping complex and highly interactive flows, but many organizations achieve benefits by simply allowing front-line staff to analyse their process flows.

Figure 36 is a pro forma which aids the listing of process flows and their analysis in order to determine possible improvements.

This technique allows the staff involved to identify non-value-adding activities and to identify the time spent on the functions shown in Figure 37.

## P is for Process management and performance measurement

To discover opportunities for improvement, it is important to review the interaction of key business processes and systems. Many of these processes will span interdepartmental boundaries and the exercise in itself will provide

# PROCESS FLOW ANALYSIS

<table>
<tr><td colspan="7">SUMMARY</td></tr>
<tr><td></td><td colspan="2">PRESENT</td><td colspan="2">PROPOSED</td><td colspan="2">DIFFERENCE</td></tr>
<tr><td></td><td>NO.</td><td>TIME</td><td>NO.</td><td>TIME</td><td>NO.</td><td>TIME</td></tr>
<tr><td>○ OPERATIONS</td><td></td><td></td><td></td><td></td><td></td><td></td></tr>
<tr><td>⇨ TRANSPORTATION</td><td></td><td></td><td></td><td></td><td></td><td></td></tr>
<tr><td>☐ INSPECTIONS</td><td></td><td></td><td></td><td></td><td></td><td></td></tr>
<tr><td>D DELAYS</td><td></td><td></td><td></td><td></td><td></td><td></td></tr>
<tr><td>▽ STORAGES</td><td></td><td></td><td></td><td></td><td></td><td></td></tr>
<tr><td>DISTANCE TRAVELLED</td><td></td><td></td><td></td><td></td><td></td><td></td></tr>
</table>

NO. _____

PAGE _____ OF _____

JOB _____

☐ MAN OR ☐ MATERIAL _____

CHART BEGINS _____

CHART ENDS _____

CHARTED BY _____ DATE _____

| DETAILS OF (PROPOSED) METHODS | OPERATION / TRANSPORT / INSPECTION / DELAY / STORAGE | DISTANCE IN FEET | QUANTITY | TIME | ELIMINATE | COMBINE | SEQUE | PLACE | PERSON | IMPROVE | NOTES (Analysis) (Why, When, Where, How) |
|---|---|---|---|---|---|---|---|---|---|---|---|
| 1 | ○⇨☐D▽ | | | | | | | | | | |
| 2 | ○⇨☐D▽ | | | | | | | | | | |
| 3 | ○⇨☐D▽ | | | | | | | | | | |
| 4 | ○⇨☐D▽ | | | | | | | | | | |
| 5 | ○⇨☐D▽ | | | | | | | | | | |
| 6 | ○⇨☐D▽ | | | | | | | | | | |
| 7 | ○⇨☐D▽ | | | | | | | | | | |
| 8 | ○⇨☐D▽ | | | | | | | | | | |
| 9 | ○⇨☐D▽ | | | | | | | | | | |
| 10 | ○⇨☐D▽ | | | | | | | | | | |
| 11 | ○⇨☐D▽ | | | | | | | | | | |
| 12 | ○⇨☐D▽ | | | | | | | | | | |
| 13 | ○⇨☐D▽ | | | | | | | | | | |
| 14 | ○⇨☐D▽ | | | | | | | | | | |
| 15 | ○⇨∩D▽ | | | | | | | | | | |
| 16 | ○⇨☐D▽ | | | | | | | | | | |
| 17 | ○⇨☐D▽ | | | | | | | | | | |
| 18 | ○⇨☐D▽ | | | | | | | | | | |
| 19 | ○⇨☐D▽ | | | | | | | | | | |
| 20 | ○⇨☐D▽ | | | | | | | | | | |
| 21 | ○⇨☐D▽ | | | | | | | | | | |
| 22 | ○⇨☐D▽ | | | | | | | | | | |
| 23 | ○⇨☐D▽ | | | | | | | | | | |

**Figure 36**  Process flow analysis chart.

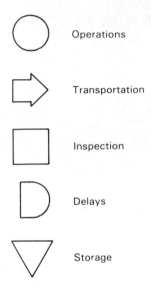

Operations

Transportation

Inspection

Delays

Storage

**Figure 37**   Symbols for showing non-value-adding activities in flow charting.

opportunity for your staff to understand more fully the needs of your business and the role they have to play in its success.

How can your work groups/departments determine how well they are meeting the needs of their internal customers, whether they are improving and whether there is a need for further improvement? To give each work group ongoing feedback on these issues, it is necessary to establish performance measures which reflect the quality of the outputs required from each group, and allow them to measure their key processes.

## Setting performance measures and analysing processes

The following is *not* a set of hard and fast rules which must be religiously followed, but is one approach to establishing performance measures within a group, and with internal customers. The technique looks at the following points:

- who your customers are
- what they need
- what their expectations and measures are

- what products and services individuals are responsible for
- what process you use to provide those products or services
- what actions you need to undertake to improve those processes.

A nine-point plan and three forms are typically used to answer these questions. The forms are listed below.

### *A team function matrix*

This document is to help you analyse a work group's function and outputs, key customers and the requirements of those customers. See Figure 38.

### *A process analysis brainstorm chart*

A chart to allow you to identify the process by which you satisfy your customers. See Figure 39.

### *A team indicator chart*

A chart to plot a work group's performance. See Figure 40.

The following nine steps are instructions to be given to those completing the documents.

## Step 1 – produce an aim and mission statement on the team function matrix

This is a sentence that defines the fundamental reason for the existence of your work group. It will be set and agreed with your work group's manager.

If one has not been given, write down your ideas and record them on the team function matrix.

You may find the following checklist useful:

1 Does the statement define the main purpose for which the work group exists?
2 Is it focused on the end objective rather than the means of achieving it? (It should not contain lengthy statements on the 'why' and 'how' of achieving the aims.)
3 Are there any parts of your statement that really belong to another group?
4 Has it been agreed with the next higher manager?

# TEAM FUNCTION MATRIX

| WORK GROUP | | | DATE: |
|---|---|---|---|
| **AIMS AND MISSION STATEMENTS** | | | |
| **OBJECTIVES** | | | |

| KEY OUTPUTS | TOP 3 CUSTOMERS | CUSTOMER REQUIREMENTS & SPECIFICATION | KEY INDICATORS |
|---|---|---|---|
| 1. | A | | |
| | B | | |
| | C | | |
| 2. | A | | |
| | B | | |
| | C | | |
| 3. | A | | |
| | B | | |
| | C | | |
| 4. | A | | |
| | B | | |
| | C | | |
| 5. | A | | |
| | B | | |
| | C | | |

**Figure 38**   Team function matrix.

## Step 2 – list your group's overall objectives on the team function matrix

These are one or more objective statements which define what your work group is trying to achieve by use of the process.

Consider the following checklist:

1  Do the statements cover all the key elements which contribute to your aim and mission statement?
2  Are the statements oriented towards your customer (either internal or external)?

## Step 3 – analyse and define the process

Use the process analysis and brainstorm chart to identify the key process and the process owner; that is, who is responsible for ensuring that the process will convert the inputs into the necessary outputs. Then go on to identify those inputs and outputs.

Have you a clearly defined method that tells everybody how to perform this work process?

If your work process is not defined, follow this checklist:

1  Identify the steps in the work process, to deliver the output to the customer at the agreed quality level.
2  Identify who has the responsibilities to ensure the process is carried out satisfactorily.
3  Identify who is the overall process owner. Who is responsible?
4  Record these decisions in a process chart or procedure.

## Step 4 – determine key outputs

Record your top outputs on the team function matrix. Remember your key input needs (as a customer yourself) for discussion with your supplier.

When recording your key outputs ask yourself:

1  Is the output described clearly, so that there is no doubt about what is being produced?
2  If you were the customer for the output, would you find the description acceptable?
3  Is your work group really the supplier for this output?
4  Are there any 'non-value-adding' outputs?

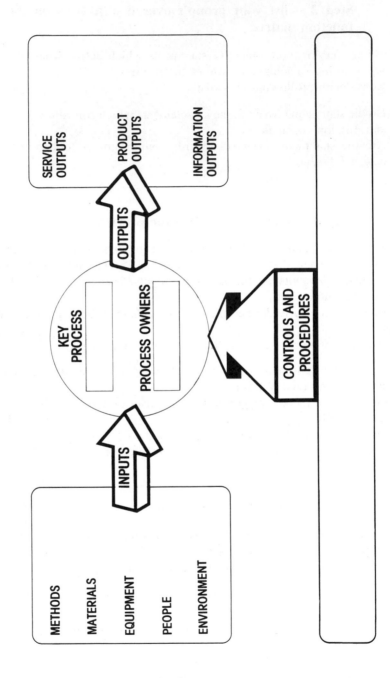

**Figure 39** Process analysis and brainstorm chart.

### Step 5 – identify key customers

For *each* of your key outputs, identify the most important customers (up to three).
   Check the following:

1  If multiple customers have been identified for this output, have they agreed that they are all customers?
2  Has the end-user been identified where your 'output' is incorporated by other work groups for another customer down the line?
3  Does your customer feel that you are the appropriate supplier?
4  Have you confirmed your role with the customer?

### Step 6 – agree customer requirements

Involve a nominated representative from the customer work groups to identify their key requirements and record them on the team function matrix. Your customers may choose to make presentations to you as to their needs and you, in turn, may do the same with your suppliers.
   Ask yourself:

1  Did the customers themselves define these requirements? If not, have you, as the supplier, verified them with your customers?
2  At this point, are you confident that the customer requirements can be met?

### Step 7 – identifying measurements of output or key indicators

For each of your key outputs identify up to three key indicators. Key indicators are measures of output that you will use to target your improvements. Record these measures on the team function matrix.
   Ask yourself:

1  Have measurements to determine the quality level been selected?
2  In general, will the selected measurements provide early indications of any possible problems or errors?
3  If you were the customer, would you be satisfied that these measurements ensure the quality of the output?

### Step 8 – collecting data and identifying improvement opportunities

Monitor your key indicators and plot the data on the team indicator chart.

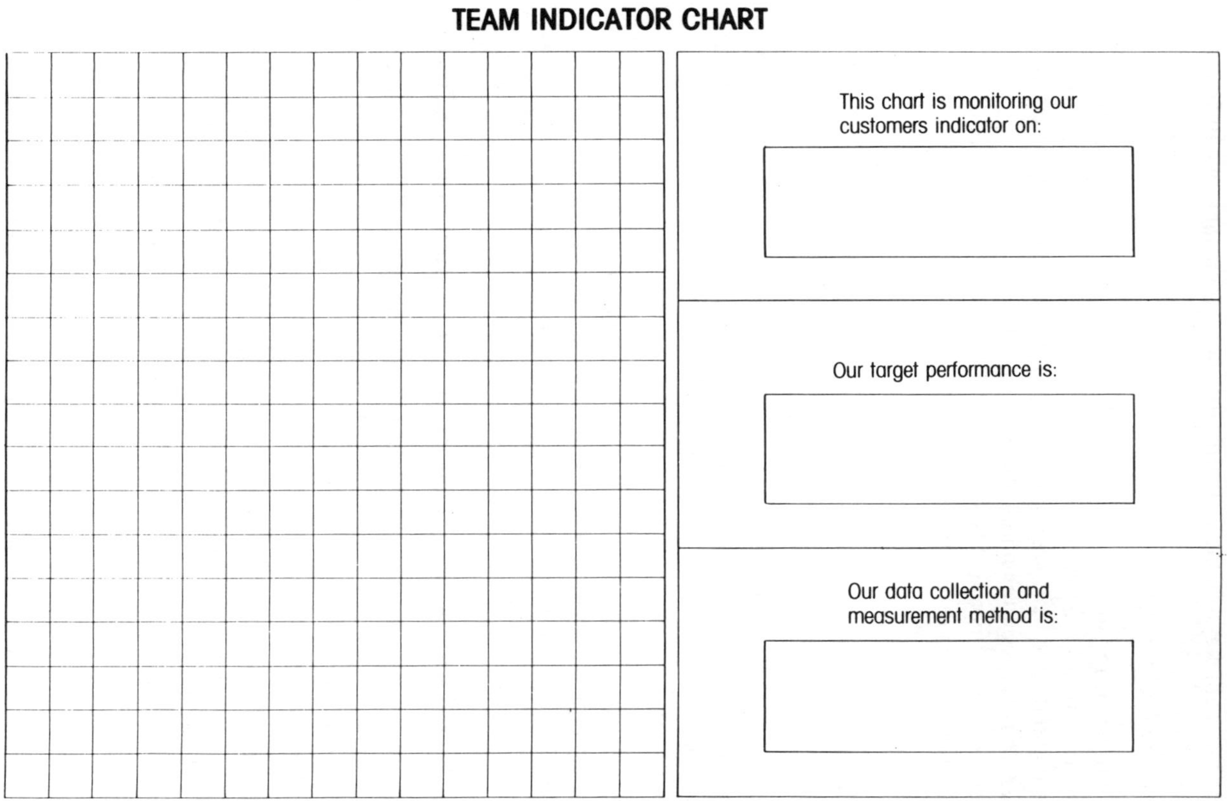

# TEAM INDICATOR CHART

This chart is monitoring our customers indicator on:

Our target performance is:

Our data collection and measurement method is:

**Figure 40** Team indicator chart.

This gives members of the work group essential feedback on performance. The following questions should be considered:

1 Is there a shortfall between 'actual' and 'target' quality levels when measuring output?
2 Do we know the cost of conformance and the cost of non-conformance?
3 Is there a problem creating a barrier to improvement?
4 Can we identify the most likely opportunity for improvement?

## Step 9 – continually measure customer satisfaction

1 Have the key measures of customer satisfaction (or dissatisfaction) been identified?
2 Have you asked your customers (or a sample of them if there are many customers receiving the same output) whether they are satisfied with your output(s)?
3 Have your customers' requirements changed (reflecting a need to change your suppliers' specification)?
4 How frequently have you planned to measure customer satisfaction?

## Possible and typical performance measures by function

### Sales

- Errors in orders
- Quotation errors
- Bid dates missed
- Change processing errors
- Customer rating

### Engineering

- Change notes
- Drawing errors/specification errors
- Audit findings: customer
                          internal
- Scrap           ⎫
- Rework          ⎬ Engineering responsibility
- Warranty        ⎭

- Design review findings and response
- Qualification test results

## Purchasing

- % of purchase reject: by supplier
                        by buyer
- Errors/omissions from purchase order
- Overdue corrective actions from suppliers: audit findings
                                            problems with product

## Manufacturing

- Defects per unit – in-process and outgoing
- % defective
- Scrap          ⎫
- Rework         ⎬ Manufacturing responsibility
- Warranty       ⎭
- Corrective action response
- Audit findings

## Field engineering/service

- Warranty cost
- Repeat failures
- % of on time completions
- Customer service rating
- Spares turn-round

## Finance

- Ledger errors
- Accounts receivable overdue
- Audit findings and responses
- Invoices not paid (by number as well as days outstanding)
- Wages errors.

For service-oriented organizations many of the performance measures will relate directly to the external customer. Measures may be established under two headings:

quantitive  –  waiting time, availability, delivery time, process time, accuracy of service, completeness of service and accuracy of billing

qualitative  –  credibility, accessibility, security, responsiveness, courtesy, comfort, aesthetics of environment and hygiene.

# Q is for Quality costs

## Quality cost philosophies

A traditional view of quality dictates that if you want better quality, then you will have to pay for it. This view holds true if quality improvement is achieved merely by increasing inspection and test activity to detect failures and then screen them out before they reach the customer. In the past a lot of effort has gone into attempting to prove that there is an optimal level beyond which there is no point in spending to improve quality. This approach misses the fundamental point that gradual continuous quality improvement should be the aim of any organization. It also does not take account of the fact that if a competitor achieves quality improvement in a particular area, then the potential loss of market share changes all previous calculations.

As Total Quality Management has evolved, a more radical approach to the cost of quality has developed. This model is used to determine the profile of quality costs within an organization against three main categories; prevention, appraisal and failure. These categories can be broadly defined as follows:

Prevention  –  these are the costs of activities that prevent failure from occurring

Appraisal  –  these are the costs incurred to determine conformance with quality standards

Failure  –  this can be subdivided into failure up to delivery or after delivery to the customer

Internal failure – these are the costs of correcting products or services which do not meet quality standards prior to delivery to customers

External failure – these are the costs of correcting products and services after delivery to the customer.

Remember that many failure costs are hidden (see Figure 41).

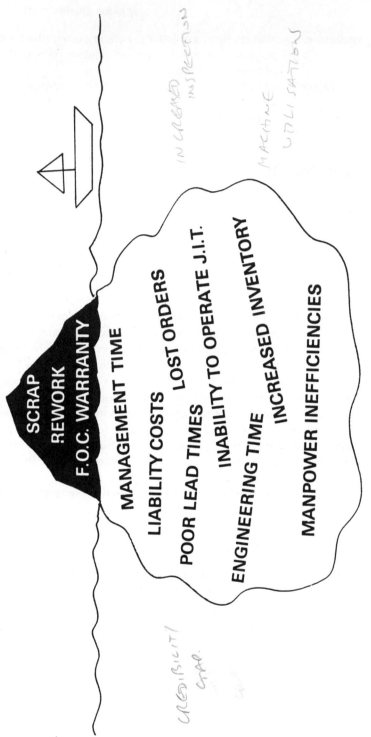

**Figure 41** The costs of quality 'failure' are mostly hidden from conventional accounting.

If a preventive culture is developed within an organization, then an investment in prevention costs offers a return in total quality cost reduction, as shown in Figure 42.

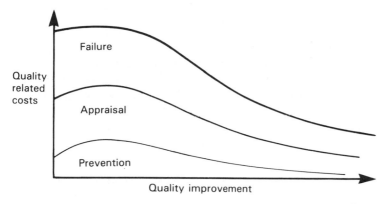

**Figure 42**   Investment in prevention reduces Total Quality costs.

In gathering quality cost data you should not lose sight of the reasons for doing so. Collating quality costs is not an end in itself – just to find out what the figures are – but a means to an end. The three main reasons for collecting and publishing quality costs are:

1  to determine the profile of the three elements and demonstrate to people whether the balance needs to shift and, in particular, what action is required if the necessary changes are to be implemented
2  to identify areas for quality improvement where action can be initiated
3  to demonstrate the impact of a quality improvement programme to develop and sustain its momentum.

### Process quality costs

You can also look at quality costs from a process viewpoint and analyse the 'costs of conformance' and the 'costs of non-conformance'.

### Costs of conformance (COC)

The costs of conformance are the costs of operating the process as specified in a 100% effective manner. This does not imply that it is an efficient, or even necessary, process, but rather that the process, when operated within its specified procedures, cannot be achieved at a lower cost. These are the minimum costs for the process as specified.

## Cost of non-conformance (CONC)

The cost of non-conformance is the cost of inefficiency within the specified process; for instance, the cost of over-resourcing or excess costs of people, materials and equipment arising from unsatisfactory inputs, errors made, rejected outputs and various other modes of waste. These are considered non-essential process costs.

Both areas of cost offer opportunities for improvement. For example, the purchasing process has cost areas as shown in Table 5.

*Table 5.* An example of process quality costs for the purchasing process in an organization

| Cost of conformance | Cost of non-conformance |
| --- | --- |
| Placing of purchase orders. | Amendments to purchase orders. |
| Normal receipt inspection. | Increased receipt inspection and labour time on rejects. |
| Inspection and test equipment depreciation and calibration. | Raising reject notes and quarantine costs (holding and administration). |
| Supplier records, performance monitoring and approved supplier lists. | |
| First issues from stores. | Subsequent issues due to shortages, changes, errors and defects. |

The listings in Table 5 are not intended to be exhaustive, but indicators of the type of costs and their classification.

## Gathering, collating and publishing quality costs

While useful guidelines can be obtained from BS6143, *Guide to the Determination and Use of Quality Related Costs*, each organization must develop its own terminology, categorization and rules for gathering, collating and publishing its quality costs. The system will be unique to that company and will therefore make comparisons with other companies difficult.

While many companies do publish their successes externally, there is no competition to see whose quality cost figures are the best; the real competition is about customer satisfaction. Therefore quality costing must be seen as a very useful tool inside the company for achieving continuous quality improvement.

The gathering of quality cost data and the acceptance of the findings within the organization is an exercise fraught with difficulties. Many of the problems can be alleviated by ensuring that the finance function plays a leading role in the exercise. In the first instance, many of the data required are available through the normal financial reporting system. Secondly, more credence is given to financial reports published by this function than to those where the authorship is seen to lie elsewhere. The finance function can ensure the quality costs are reported against the same base as other financial reporting to management.

In many cases the quality costs will not be readily available from normal accounts. Examples are the costs of time spent correcting errors on invoices, management/supervisory time spent on preventive activities versus firefighting and so on. In the first case, sampling the activity will determine an average cost and the number of occasions it arises. The second example can be approached by using periodic surveys asking staff to record how they apportion their time to various activities. Although the use of such sampling and survey techniques obviously produces only approximations, it is much more cost effective to derive these synthetic data than to spend excessive time trying to pin everything down to the last penny.

Remember that quality costs are used to highlight areas for improvement and then monitor success. Therefore, it is not the absolute accuracy of any set of figures which is important, but the difference between one point in time and the subsequent reporting.

See also P is for Problem solving.


# Q is for Quality definitions


### A definition of quality

There is a saying that the man who invents another word for 'quality' will make himself very rich. Until that time many definitions for quality, quality assurance, quality control, quality management and, particularly, Total Quality Management will continue to exist. Certain national and international standards organizations have attempted, and in some cases achieved, their own definition of certain terms, but the precise definition of a 'Total Quality process' will always be in the hands of an individual or an organization. A few of the internationally accepted standard definitions are given below (refer also to BS4778: 1987 (ISO8402: 1987), *Quality Vocabulary Part 1, International Terms*) but remember that quality is riddled with 'jargon'. You must create and communicate in your own organization a 'common language' as the only jargon that matters is the terminology that your staff and colleagues will understand.

Here are a few definitions of quality:

> 'Meeting or exceeding customers' expectations at a price that represents value to them' – Harrington.

> 'Fitness for purpose of use' – Juran.

> 'The totality of features and characteristics of a product or service that bear on its ability to satisfy stated or implied needs' – BS4778: 1987 (ISO8402: 1987, *Quality Vocabulary Part 1, International Terms*).

> 'The total composite product and service characteristics of marketing, engineering, manufacture and maintenance through which the product and service in use will meet the expectation by the customer' – Feigenbaum.

Some more precise terms and descriptions are gaining international acceptance:

> Quality management: that aspect of the overall management function that determines and implements the quality policy.

> Quality policy: the overall quality intentions and direction of an organization as regards quality, as formally expressed by top management.

> Quality assurance: all those planned and systematic actions necessary to provide adequate confidence that a product or service will satisfy given requirements for quality.

> Quality control: the operational techniques and activities that are used to fulfil requirements for quality.

> Quality system: the organizational structure, responsibilities, procedures, processes and resources for implementing quality management.

> Quality plan: a document setting out the specific quality practices, resources and sequence of activities relevant to a particular product, service, contract or project.

> Quality audit: a systematic and independent examination to determine whether quality activities and related results comply with planned arrangements and whether these arrangements are implemented effectively and are suitable to achieve objectives.

# Q is for Quality function deployment

Quality function deployment (QFD) is a technique whose origins can be traced back to Japan in 1972. It has reached Europe via its adoption in certain sectors of US industry. Its foothold in the USA started early in 1986 in the automotive industry.

## What is QFD?

QFD is a powerful tool for use within a Total Quality Management programme.

Its particular power derives from the fact that it is a planning tool for use upstream, off-line, to ensure that the customer's needs are first understood in the customer's own terms; then deployed into design requirements and subsequently through the manufacturing chain of critical part characteristics and key process requirements; and finally deployed to operational specifications. See Figure 43.

When properly applied, for example, what the operational staff are machining, moulding, assembling and so on will have a continuous common thread which is traceable back throughout the organization to what the customer really wants, to an optimum quality level.

## Understanding what the customer wants

The voice of the customer is the cornerstone of the QFD process. There are many ways of determining what this voice is and it is preferable to use diverse meaningful information. Thus customer surveys, warranty information, customer clinics, dealer information, competitive surveys and so on will all contribute to the final list of customer requirements.

## Translating customer wants into manufacturing specifications

How customer wants are transferred to design requirements is not always as straightforward as it might seem. For example, a consumer will not expect the rain to penetrate into the car interior by way of the closed door, but will also have a requirement that the door may not be unduly difficult to close.

In this case the provision of a weatherseal will keep out the rain, but it will also affect the door-closing effort. Judging the importance of these two requirements and how they impact on one another is a fundamental part of the QFD methodology. By depicting on a chart how each particular want is to be met at right angles to the want itself, the complexity of interdependencies is simplified. Additionally, the strength of the relationship of each want with all the 'hows' may be objectively assessed in four categories – none, weak, medium and strong. Figure 44 shows this for car door and resin penetration examples.

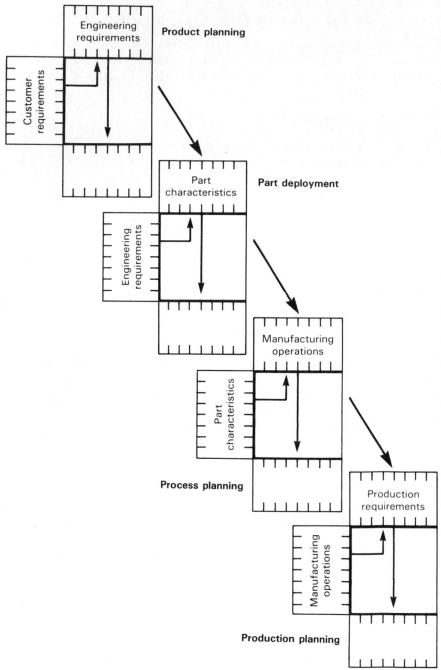

**Figure 43** QFD ensures that customer needs are planned through design and production. (Redrawn courtesy of the American Supplier Institute, Inc.)

**Deploying the 'voice of the customer'**

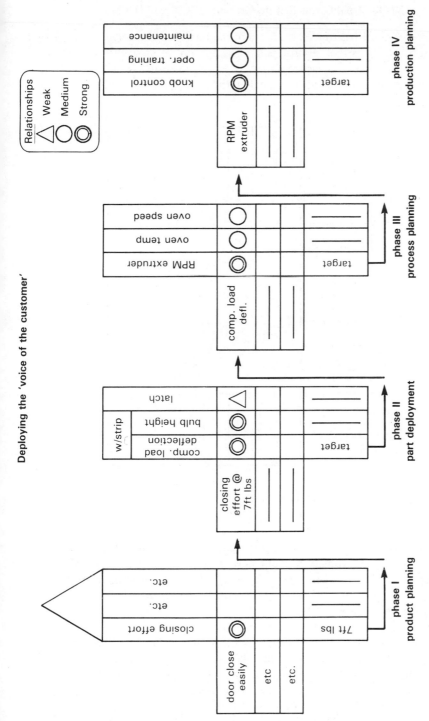

**Figure 44** Relating a customer need to the required process controls. (Redrawn courtesy of the American Supplier Institute, Inc.)

Having identified the various relationships that exist between the design requirements, it is then appropriate to assign target values to each requirement – the 'how much' shown on the chart in Figure 45. These will normally be nominal dimensions, or the smaller the better, or the bigger the better.

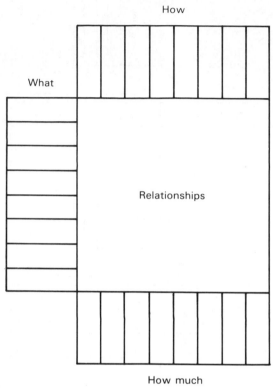

**Figure 45**  Assigning targets. (Redrawn courtesy of the American Supplier Institute, Inc.)

At the same time, each customer want can be graded on a scale of 5 to 1, so that the importance of each 'how' may be assessed more objectively in relationship with other 'hows' by multiplying the grading with the relationship rating – none, 0; weak, 1; medium, 3; strong, 9. Thus column 1 of Figure 46 totals $(5 \times 3) + (2 \times 9) = 33$.

Future areas of conflict may be avoided by considering the relationship between each of the design requirements. For example, there is a negative relationship between the desire to have a strong thick section of weatherstrip and the desire to have minimum compression force of the section in order to facilitate the characteristic of minimum closing effort of the car door. These relationships are entered in the 'roof' of what is

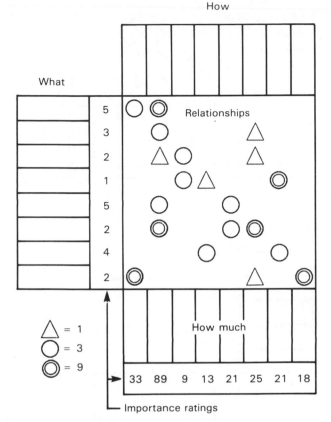

**Figure 46** Grading customer needs. (Redrawn courtesy of the American Supplier Institute, Inc.)

sometimes referred to as the 'house of quality'. This is shown in Figure 47.

In product design in particular it is usual to **benchmark** competitive products. The QFD chart is a convenient device to check how well the various identified customer wants have been met in those products benchmarked. It is then useful and often illuminating to see to what extent a well or badly considered customer assessment agrees with the engineering technical assessment for the feature. These assessments are shown respectively opposite the customer wants and below the 'how much' table. See Figure 48.

Figure 48 shows the whole concept of the QFD chart at a first phase of product planning. The format of the chart is not sacrosanct, but should embody at least the principles outlined above. Figure 49 shows an 'important control' panel at the bottom of the chart. Typically, government regulations affecting design features may also be shown. This example is for a car door design.

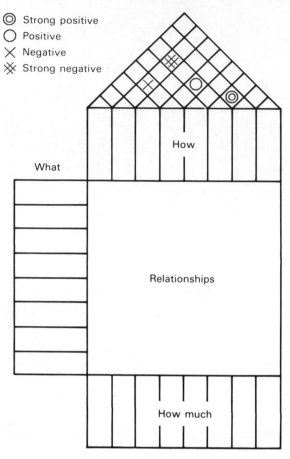

**Figure 47**    Correlating the relationship between desired characteristics. (Redrawn courtesy of the American Supplier Institute, Inc.)

A QFD chart is a living document of the output of a multidisciplined team working through the process. The information collated is to be used in the pursuit of a product with highly perceived quality, achieved in minimum development time and manufactured right first time at minimum cost.

## The benefits of QFD

The benefits of QFD, if properly applied, may be summarized as:

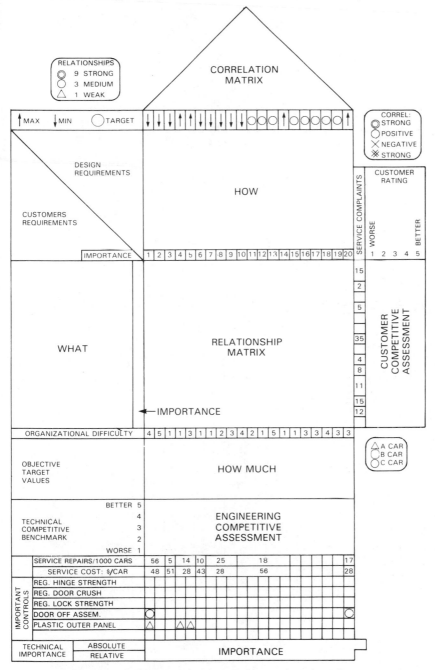

**Figure 48** QFD can include customer and engineering assessments of the competition. (Redrawn courtesy of the American Supplier Institute, Inc.)

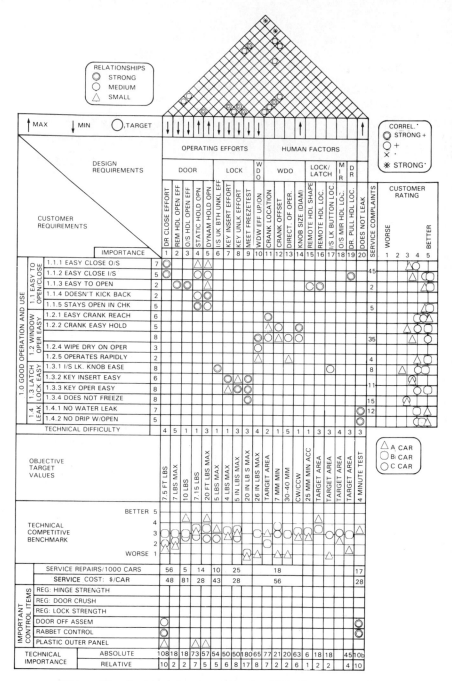

**Figure 49** A complete example of a QFD chart assessing a car door design. (Redrawn courtesy of the American Supplier Institute, Inc.)

Accountable benefits:

Increased customer satisfaction
Reduced development time
Reduced late engineering changes
Fewer start-up problems
More even production flow
Reduced start-up costs
Customer-driven integrated system of material and operational specifications.

Follow-on benefits:

Rational quality engineering and assurance plan
Knowledge base which applies – and can be transferred – to similar products.

# R is for Recognition and reward

## Recognizing individual contributions

Successful Total Quality programmes recognize that if everyone in the organization is to become involved in the process of managing quality improvement at their level, then it is important that their efforts are recognized in order to encourage them and maintain the momentum of the process. In the first instance, simply being listened to, possibly for the first time, by those more senior in the organization provides a very fundamental form of recognition. When their ideas and suggestions are taken on board, the feeling of being able to influence change and the way they work is a reward in the form of recognition of the importance of people to the company.

This approach should be supported by a personal expression of praise for achievements in quality improvement. This needs to be forthcoming from management at all levels through daily personal contact and through the more formal channels of communication. People will be used to having their shortcomings highlighted in one way or another. While this feedback is important for quality improvement, it needs to be balanced by praise when it is deserved. Management often find it easier to give a rap over the knuckles than a pat on the back when required.

The more formal channels of communication may include quality noticeboards, bulletins and company newsletters. These should be used to give company-wide publicity for significant efforts in achieving quality improvement. This form of recognition can include articles detailing particular projects, how they were carried out and the benefits to the company, and including the names and perhaps photographs of those involved. Some companies are now adopting the medium of video to capture their quality success stories; this can be used for subsequent training and at company conferences. The use of such media can add an extra dimension to the process of providing recognition.

## Methods of reward

Many companies are beginning to use existing events like conferences and company dinners, or in some cases establishing new ones, to highlight quality improvement and give recognition to those involved. Those involved in quality achievement and recommended by their manager or peers for recognition will attend these events as honorary guests and sometimes receive gifts, awards or tokens of recognition. The events can also be used as a forum for presenting the results of their work.

Other ways of rewarding employees may take the form of external visits which traditionally have only been open to managers at certain levels and in particular functions. The visits can be to customers' or suppliers' premises, which is a recognition of the employees being good ambassadors for the company. They also give those involved an insight into how the output of their work is used and also how the input to it is achieved. Their experience can then be brought back into the company and shared with their fellow employees, thus increasing awareness further. Some external visits, in particular attending exhibitions, are often regarded as perks. Again, employees who would not normally be given the opportunity can be involved on such occasions as a reward for quality achievement. The exposure to the company's customers and potential customers is an experience that can be usefully shared within the company.

## Recognizing group contributions

Careful consideration needs to be given to whether recognition should be for individual or group effort. In most cases quality improvement can only be achieved through the combined efforts of a group of people. If only certain individuals are recognized, this could have a demotivating effect on those who also made a contribution to any achievement. However, if an individual makes an outstanding contribution to quality improvement, it should be possible for the group to elect the individual for special recognition. In these cases any reward to individuals provides some recognition of the group to which they belong.

### Cash rewards

Finally, the question of financial reward needs to be addressed. It should be recognized that there can be dangers in providing direct cash reward for quality achievement. If the cash awarded does not meet the recipient's assessment of the worth of the quality improvement, it can have a demotivating effect. There is also a greater problem if the individual is financially rewarded and contributions by other employees are not recognized. In successful businesses the long-term financial rewards to all employees often reflect the cost savings being achieved through a Total Quality initiative.

See also E is for Employee participation.

## S is for Seven statistical tools

This term originates in the Japanese practice of using basic statistical tools and teaching them to everyone. It relates to elementary and indispensable tools for problem solving. They are applied by everyone from company presidents to line workers and are not the preserve of the quality department.

The seven tools are:

1  check or tally sheet
2  cause and effect diagram
3  Pareto diagram
4  stratification
5  scatter diagram
6  histogram
7  control charts.

The more detailed, most commonly applied tools are described separately. The others are described here.

See also P is for Problem solving, P is for Pareto analysis, C is for Cause and effect diagram and C is for Control charts.

Readers who require an introduction to basic statistics should refer to the bibliography.

### Check or tally sheet

| | | |
|---|---|---:|
| Wrong parts fitted | JHT JHT JHT JHT IIII | 24 |
| Faulty parts fitted | JHT I | 6 |
| Parts assembled incorrectly | JHT JHT JHT JHT JHT JHT | 30 |
| Parts missed out | JHT JHT II | 12 |
| Damaged parts | III | 3 |

**Figure 50**  Check or tally sheet of assembly faults.

Check or tally sheets are used to categorize items. The check or tally sheet is very straightforward. On the left-hand side are listed the categories being analysed; these might be causes of quality problems, as shown in Figure 50. Every time that a cause is highlighted as being the problem, a tally is made; the tallies are then grouped into sets of five to make assessing the final total easier. This is often the initial step in analysing a problem. The next stage is often to present this information in graphical form. (See Figure 33 under P is for Pareto analysis.)

## Histogram

The histogram is a method of representing data. In the earlier example of incorrect assemblies, it might be necessary to look further into the incorrectly assembled parts. Figure 51 shows a particular aspect, namely how often these faults occur on a daily basis.

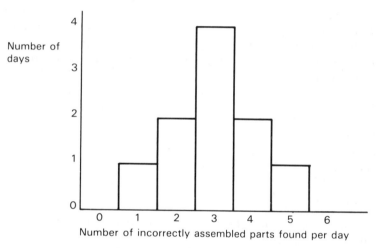

**Figure 51**   Histogram of incorrectly assembled parts.

## Stratification

Stratification can be used to analyse information further. In the example given in the histogram (Figure 51), stratification by breaking the results down into two shifts helps to identify where problems are occurring. This is shown in Figure 52.

## Scatter diagrams

Scatter diagrams are used to analyse the correlation between two variables as shown in Figure 53.

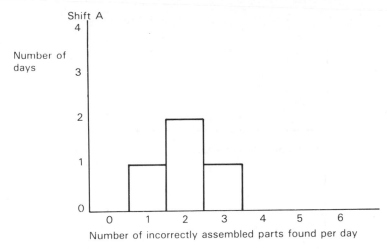

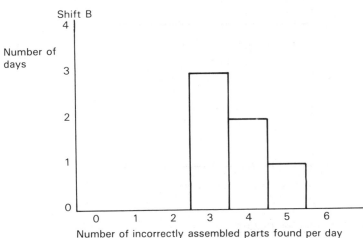

**Figure 52**   Stratification of incorrectly assembled parts.

In graph (a) an increase in variable X will give a corresponding increase in variable Y; this is positive correlation. In graph (b) an increase in variable X will bring about a decrease in variable Y; this is negative correlation. In graph (c) an increase in variable X will not affect variable Y; this is called zero correlation.

# S is for Statistical process control

### What is it?

Statistical process control (SPC) is a generic heading for a range of statistical

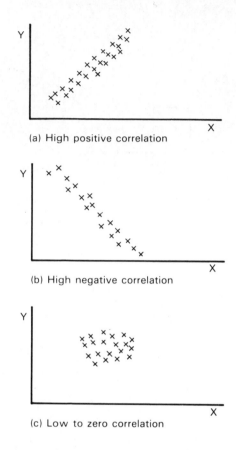

(a) High positive correlation

(b) High negative correlation

(c) Low to zero correlation

**Figure 53**   Scatter diagram.

techniques that improve the performance of a process by reducing its variability.

Despite the use of statistics in its application, SPC is often applied jointly with other non-statistical problem-solving techniques such as **brainstorming**.

Figure 54 shows how the use of statistical techniques helps process improvement. Consider the three sets of samples taken at various steps in time.

Process improvement

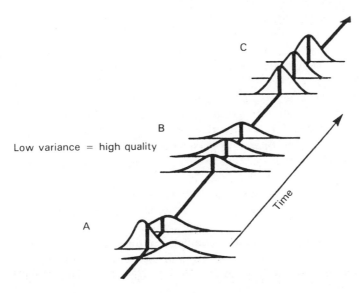

**Figure 54**   SPC can reduce variation and improve accuracy.

At time A – all samples give a different view of the process, both in spread and accuracy terms.

⇩  Improvements

At time B – all samples show a similar view of the process; it is on average accurate but has a wide variation.

⇩  Improvements

At time C – all samples show a process that is accurate with a low variation.

Because statistical process control has wide applications in manufacturing, it is usually regarded as a manufacturing control technique. However, *any* process can be analysed through SPC by selecting suitable key parameters to measure.

Examples are journey time and the number of rings that occur before a telephone is answered.

## Why use statistical process control?

No process is able to produce consecutive items which are identical. Differences will occur due to sources of variation which are inherent in the process or machine; examples are backlash, vibration, fit, speeds and power. The extent of this variation can be measured and is called the *machine capability*.

In addition to these *inherent variations*, other *special causes* – such as variation in raw material, operator performance, tool wear, machine setting and maintenance control – will influence the output. These things can generally be controlled by the operating personnel and support services.

The combination of inherent and special causes of variation generates the overall process performance.

It is important that we distinguish between inherent and special cause variation and treat them separately. Inherent variation is built into the system and can only be reduced by management action.

Special causes of variation generally occur on an irregular basis; examples are broken or worn tools, faulty material and poor machine setting. Often, although not always, these are local faults which can be corrected by the process operator.

In the past, manufacturing personnel were responsible for the production of goods which were inspected and sorted by quality control to ensure that no 'out of specification' parts reached the customer. Similarly, in office systems, work is often checked and rechecked in an effort to identify errors. This emphasis on the detection of errors *tolerates* waste. It wastes the manpower used to inspect good production; it wastes the manpower used to manufacture substandard products; it wastes machine capacity; it wastes material.

A more effective method of working is to avoid waste by not producing defective parts in the first place. This requires everyone to concentrate on *defect prevention* rather than *defect detection*.

SPC is a means for achieving the prevention of defects by highlighting situations when the output of the process is drifting outside acceptable limits and identifying whether the cause of that drift is due to *inherent* or *special* cause variation. It assists by providing benefits in the following areas.

### *Minimizing operational losses*

Minimizing losses is achieved by systematically identifying and analysing key processes/products and directly looking at the root causes of problems associated with them.

*Changing culture*

Incorporating statistics universally into the organization at all levels, as a tool of Total Quality, will progressively transfer the culture of the entire organization from one of defect detection to prevention. In passing, this may involve the beneficial transfer of the 'checking' role from an inspector to an operator.

*Creating team spirit and involvement*

SPC enhances the TQM requirement for involvement at employee level. Employee interests become congruent with company interests and help to create an active team environment to solve problems.

*Establishing means of monitoring individual performance*

SPC makes all employees aware of the benefits of good quality performance – providing visibility of achievement through statistical monitoring of their own processes and performance.

*Assisting internal customer communication*

SPC provides a common language between the customer, the designer and the producer of products and services, minimizing misunderstandings of customer requirements.

*Assisting external customer communication and assurance*

SPC is often a means of communicating confidence and assurance to the customer that the supplier's products and processes are in control. In certain sectors, the motor trade for instance, this assurance will be a specified requirement of the contract.

The most common approach to SPC is shown below; each of the techniques quoted is considered in more detail elsewhere:

1 Take an initial sample from the process; use statistical calculations to construct a control chart.

2 Carry out regular sampling and record a control chart to bring process 'in control'.

3 Once a process is in control, capability studies can be carried out to measure the ability of the process to achieve the required specification.

See also S is for Suppliers, C is for Control charts, C is for Capability, P is for Problem solving and S is for Seven statistical tools.

# S is for Suppliers

## Suppliers and improvement

### Suppliers as partners

The nature of the relationship that manufacturers are seeking with suppliers is changing. The traditional 'conflict of interest' approach, where communication between customer and supplier is guarded in case the information given can be turned into an advantage by one party at the expense of the other, is no longer acceptable to major purchasing organizations.

It is recognized that a closer, long-term collaborative relationship is vital to both parties for the achievement of quality objectives, desired market share and profitability. The supplier has become an essential part of any strategy aimed at improving the effectiveness of a business.

A business partnership attitude is increasingly being demanded from suppliers. Suppliers are expected to use their specialist knowledge to contribute to the development of customers' products and to consider the quality and reliability of their own products from the customer's point of view. Customers, on the other hand, are prepared to guide and assist suppliers with the improvement of their quality and productivity. Nissan Motor Manufacturing (UK) Ltd, for example, state: 'Our policy is to develop long-term relationships with a select group of suppliers' and 'Working with selected suppliers who are committed to achieving improvement, it is intended to provide direct assistance with quality and productivity improvement programmes'. The Philips Group have coined the phrase 'co-makership' to describe the working together with their suppliers towards a common goal.

### Demands for supplier quality

For successful collaboration, customers increasingly are insisting that suppliers develop a pro-active style of management which anticipates and resolves problems before they occur, and which has as its objective an ongoing improvement in quality and productivity – in short, an effective TQM strategy within the supplier's organization.

In addition to having to operate a documented quality system which complies with the customer's requirements, such as Ford Q101, suppliers in the automotive industry are expected to be conversant with the latest quality assurance techniques and to demonstrate effective use of those techniques which are appropriate to the control of the supplier's activities. For example, most major automotive manufacturers advocate the use of such techniques as **failure mode and effects analysis, statistical process control, quality function deployment, Taguchi** quality engineering and so on.

### Assessment and approval

Formal assessment of suppliers is now extended beyond the assessment of the supplier's quality system. Consideration is made of warranty claims and production problems attributable to the supplier, of the performance in respect of initial samples and, significantly, of the supplier management's commitment to the improvement of quality. Austin Rover, for example, allocate 15% of the total supplier rating marks to 'supplier management awareness and commitment to quality'.

Compliance with formal quality management requirements has been extended to tooling and equipment manufacturers, who, surprisingly in view of the critical effect of the manufacturing equipment on product quality, had not been subjected to the same strict supplier quality assurance requirements as the suppliers of parts and services. The Chrysler Corporation of the USA, for example, produced a comprehensive 'Tooling and Equipment Supplier Quality Assurance' document in January 1989.

Outside the automotive industry, there is a trend towards third-party (e.g. British Standards Institution) assessment of suppliers' quality systems to recognized quality system standards, such as **BS5750** (ISO9000). This is particularly evident in Britain, where a significant number of purchasing organizations are virtually eliminating their own appraisal of suppliers and insisting that suppliers become registered under the BS5750 scheme. A number of purchasing organizations now have a policy of dealing only with suppliers registered under a recognized quality system standard.

### Assisting and training suppliers

Customers now accept that they have a responsibility not only to provide suppliers with good quality information on time, but also to assist suppliers with their improvement programmes. Major manufacturers are establishing 'supplier development programmes' to assist key suppliers with the establishment of quality systems, with the introduction and application of techniques such as **statistical process control**, with improvement of productivity and with the training of suppliers' personnel. Training may involve a supplier's key employees working for short periods at a customer's premises to gain appreciation of the customer's needs, as encouraged by Nissan (UK), or attending training courses related to a customer's requirements which are approved by the customer; for example Ford Q101 and SPC requirements.

## Liaison with suppliers

In general, a more structured approach to collaboration with suppliers is being developed, particularly for the control of activities concerned with the introduction of new and modified products. Nissan (UK) operates a 'pre-production quality assurance' scheme which specifies the controls to be taken for the effective introduction of new products, or changes to existing products, and which fully involves suppliers. Joint problem-solving teams, which can include representatives from design, engineering, purchasing, production and quality functions, are established to identify and resolve potential problems during pre-production, and problems occurring during production.

Close collaboration with suppliers, however, does not reduce the supplier's responsibility for supplying non-conformance-free products and services. If anything, customers are now adopting a strategy of recovering from suppliers costs of non-conformance attributable to the supplier. The intention is stated clearly by Nissan: 'Nissan does not expect to receive any defective parts and so carries out very little incoming inspection. If however there are warranty claims attributable to a supplier then Nissan expects the entire cost to be borne by the supplier.'

Increasingly, suppliers are being judged by their performance in terms of commitment to the customer and their responsiveness to problem solving, in addition to the way they perform in respect of quality, price and delivery. The search for capable suppliers who are committed to the improvement of quality and productivity is already well on the way. The first indication is that a reduction of a customer's supplier base by as much as 50% may not be uncommon, being the result of discarding suppliers. Ford Europe reduced their supplier base from 2,100 to 1,300 by 1987 with the aim to reduce the number of suppliers to around 1,000.

Suppliers who can demonstrate acceptable performance and ability to collaborate successfully with their customers become 'preferred suppliers' and are assured a long-term and stable relationship. Since it is easier to develop a long-term business partnership if suppliers are in close proximity, customers are now reviewing their international sourcing strategies to develop shorter supply lines. Closeness, of course, is also a vital element in the use of just-in-time (JIT) purchasing strategy.

The extent to which suppliers may have to collaborate with their customers is illustrated by the fact that a number of major purchasing organizations are insisting on electronic data interchange (EDI) with key suppliers. The data exchange relates not only to quality and product engineering, but covers also schedules, lead times and inventory management.

The changing strategies towards suppliers stem from customer realization that in the increasingly competitive environment, over a long term, it is the cost of doing business with a supplier which is important and not the price of a piece. Having come to this conclusion, it is quite clear that the major purchasing organizations are determined to deal only with suppliers who are prepared to collaborate towards the improvement of quality and productivity to the benefit of both parties.

## Your TQM strategies for suppliers

To summarize, any Total Quality programme reaches outwards at both ends of your business, towards your customers and your suppliers. The approach you take to your suppliers will be governed by the nature of your supplier base and your business volumes. The points highlighted below are some of the 'keys' to your supplier strategies.

### Joint design reviews

If you are involved in design work you should consider involving represent-atives from your suppliers as part of the team. Not only will this help to forge closer relationships, allowing them to respond more readily to your needs, but it will also add value to your design process.

### Supplier seminars

As you develop your own quality initiative, you may find it appropriate to expose your suppliers to the cultural change which your company has undergone. This will help your suppliers to identify the changes they need to make to form a successful partnership with your company in the future. Some companies use such occasions to indicate the intention of reducing their supplier base, with only those prepared to make the change being part of future plans.

### System requirements

A first step for many suppliers may be to adopt similar quality system standards to those adopted by your company. Doing so offers assurance that they are planning to meet your quality requirements, or at least provide a basis for their assessment.

### Statistical process control

Encourage suppliers to adopt such statistical techniques as may provide suitable evidence of control, to allow you to reduce and possibly eliminate goods receiving inspection.

### Long-term contracts

If suppliers are being asked to institute changes and develop an approach in line with that of your company, then a commitment to a longer-term relationship may need to be demonstrated. Some companies are offering inducements in the form of increasing the length of contracts in line with improvements in supplier performance.

### Bonuses and penalties

A variation on the above approach is the use of 'carrot and stick' methods, offering premiums for consistent performance and asking suppliers to demonstrate their confidence in their ability by committing themselves to penalty clauses. While these two extremes may either be unacceptable or difficult to enforce, the middle road of offering prompt or early payment for reliable performance may help to obtain the commitment of suppliers.

### Product or system audits

Part of your strategy to gain assurance and perhaps influence your suppliers' methods is to carry out audits on their premises. Auditing their systems will enable you to develop their methods in line with your requirements. Similarly auditing products is one step used by companies towards a just-in-time approach.

*Vendor ratings*

Giving your suppliers sufficient feedback on their performance is important in encouraging quality improvements. If this information is being generated, then it should be made available to your suppliers.

*Paperwork*

Establishing a common approach to paperwork may help to improve communications between your company and its suppliers. With the technological advances available, some companies are establishing electronic links with their suppliers to speed up communications.

### Your performance for suppliers

Whether you integrate these points into your strategies or not, most businesses would benefit from closer relationships with their suppliers. When was the last time you asked your supplier to comment on how well *you* perform?

See also Q is for Quality function deployment and S is for Statistical process control.

# T is for Teamwork and teambuilding

### Teamwork

The participative approach is characterized by several types of employee groups. These are variously known as quality circles, quality improvement teams, continuous improvement teams and employee involvement groups. In all of these, groups of employees are trained in problem-solving techniques with quality improvement in mind.

Improvements are achieved through projects that the teams carry out and by integration of the teamworking philosophy into everyone's day-to-day job. The elements of this philosophy are as follows:

- The team should have an objective to develop the team members' individual abilities as well as to make improvements for the organization's benefit.

- In solving problems and making improvements, team members should help each other to learn and develop, thus encouraging good team spirit.

- Creativity and innovation must be encouraged; therefore no one person should be allowed to dominate and criticism within the team must be avoided.

- Employees should not be forced to participate in teams, but should be encouraged and persuaded with regard to the advantages both for them and the company.

- All team members should be trained in problem-solving techniques as well as being presented with a structured approach to improvement.

- The support and involvement of management is essential. They should not feel threatened by the team or take credit for team success. Trust and co-operation should be enhanced; management should recognize and promote team success, taking care that the focus is on the group, not on individuals. It is vital that management provide the time and resources for the teams to operate effectively.

Teams usually consist of between three and twelve volunteers who share the same area of responsibility; this may be at any level in an organization. The team will include a leader and facilitator, the leader usually, but not necessarily, being the area supervisor.

## The role of the facilitator

The role of the facilitator is to support the leader and to act as a co-ordinator and administrator for training, access to technical expertise, and presentation and communication of results.

Various task roles and people roles can be identified within a team. It is important that these roles are recognized so that possible effects on the team can be identified and also so that the roles can be used to the best advantage of the group. This task is just one of the responsibilities of the team leader and facilitator.

It is important, particularly at the very early stages of team development, that the group has a strong, effective leader who keeps a balance between the roles of manager, trainer, educator and motivator. Groups recognize and appreciate good leadership, the qualities most often associated with it being competence and job knowledge, honesty and fairness, motivity and emotional calm, trust and openness, and friendliness and approachability. Although a lot of these leadership skills may be inherent in someone's behaviour, specific elements can be acquired.

### Teambuilding

The team leader and facilitator should be concerned to develop the group so that it works effectively as a team. One of the first steps in teambuilding is to ensure the team develops and agrees on its goals and objectives. Milestones should be discussed and communicated. When they are reached the team can celebrate success, thus building team spirit.

The team should be encouraged to talk in terms of 'us' and 'we' to establish a unity and could even give itself a group name or logo. Teams may also be encouraged to develop rituals and ceremonies.

Team meetings should be held for a maximum of an hour per week to begin with, dropping to perhaps one meeting per fortnight once established. To begin with, the meetings will involve training of team members, but as they become proficient, problem analysis and solution will be the main objective.

See also E is for Employee participation and O is for Organization and structure.

# V is for Visual control

An important aspect of **communication** is that it should enable people to know what is happening in their work area as soon as possible after it has occurred. This can be viewed in two ways.

1  When an abnormal condition exists, corrective action needs to be taken immediately. There needs to be a signal as soon as this event takes place. Various systems can be used; for instance, if an operation has stopped or a poka-yoke device has detected a defect item, a prominent light may indicate the fact.
2  Recognition and information feedback should be given to the work area as soon as possible. Display boards around the workplace not only develop a visual and open management system which keeps the workforce aware, but also give a self-explanatory information system which shows visitors just how serious the management are about quality. Typical examples of information which can be displayed are:

  - **control charts**
  - standard work sheets
  - maintenance sheets
  - safety check points
  - map of **poka-yoke** devices
  - recognition for improvements
  - housekeeping evaluation
  - improvement activity status

- defect levels – targets and status
- set-up charts
- skill development charts.

Whilst there is considerable potential in introducing an open style of information sharing, it is important that there is consideration of the company culture and people's feelings.

In summary, the objectives of visual control are:

1 enhance communication and participation
2 control quality process at source.


# W is for Where does it apply?

It is a common misconception that Total Quality can be applied only to those organizations where a product is manufactured. That is wrong: Total Quality is just as relevant to organizations providing a service; in fact, it is debatable whether there are companies that are wholly one or the other.

First, we must ask whether there are companies which provide only products and others who provide only services. For both types of industry a balance of both product and service is provided, but the balance is different from company to company. When you buy a car, the transaction is largely about a product with some element of service during and after sale. Having a meal in a restaurant is an experience probably split 50:50 between product and service. Taking a train journey is predominantly buying a service but minor product elements such as tickets, timetable and buffet car food will be involved.


## Applying Total Quality to service organizations

The need for quality in a company manufacturing a product is obvious and a great deal has been written which concentrates on those types of organization. Let us consider, for a moment, how quality affects the service sector and then take an example from both product and service sectors to try to highlight how Total Quality has been applied.

To some extent managing service quality is more difficult because some elements are intangible and difficult to measure. However, one advantage can be that a greater percentage of a company's employees are in contact with the external customer; therefore customer orientation can be easier to achieve.

This increased customer contact has been described as 'moments of truth' by Jan Carlzon, president of the Scandinavian Airlines System (SAS). He says that each time customers come into contact with an organization they form an impression; each time is a moment of truth. The sum of the moments of truth forms the overall company quality rating. To be successful, companies need to manage the moments of truth so that a good quality impression is formed and the customers will come back.

The following are some commonly held perceptions of what makes a service different from a product. Do you agree? Do any of these apply to your company?

1 Sales, production and consumption of a service take place almost simultaneously.

2 A service cannot be centrally produced, inspected, stockpiled or ware-housed. It is usually delivered to wherever the customer is, by people who are beyond the immediate influence of management.

3 A service cannot be demonstrated, nor can a sample be sent for customer approval or trial in advance of purchase. The provider can explain, promise and tell how the service has benefited others, but the service doesn't exist for the prospective customer.

4 The person receiving the service generally owns nothing tangible once the service has been delivered. The value is frequently internal to the customer.

5 A service is frequently an experience that cannot be shared, passed around or given away to someone else once it is delivered.

6 The more people there are involved in the delivery of the service, the less likely it is the receiver will be satisfied.

7 The receiver of the service frequently has a critical role to perform in the actualization of the service. The customers, then, have to know their role in the delivery process.

8 Delivery of the service frequently requires some degree of human contact. Receiver and deliverer frequently come into contact in some relatively personal way.

9 The receivers' expectations are critical to their satisfaction with the service. What they get, compared or contrasted to what they expect to get, determines satisfaction.

10 Exerting quality control over a service requires monitoring of processes and attitudes.

## Total Quality in the UK National Health Service

Activity currently being undertaken in the National Health Service (NHS) will now be used to show the potential influences of Total Quality in the service sector.

The NHS is one of the largest organizations in Europe with over a million employees and a 'turnover' of £3.5 billion on goods and services. The Secretary of State for Health is responsible for the 14 regional health authorities and 190 district health authorities in England, operating through 2 central management and supervisory boards. The health authorities in Scotland, Wales and Northern Ireland are the responsibilities of their respective secretaries of state. The quality activities and requirements of these organizations have had a signficant effect on many UK industrial sectors.

Each of the health authorities is responsible for procuring materials and services. The central procurement directorate has the role of providing advice in terms of quality and reliability, and of the logistics of ensuring effective and economical supply. The Department of Health (DoH) set up a manufacturer registration scheme in 1981 to ensure that medical products were both safe and fit for purpose. This originally covered sterile products and cardiac pace-makers, but has been gradually extended to include powered medical equipment, orthopaedic implants and rehabilitation equipment. It was hoped that the scope of the scheme would eventually be widened to cover all health care equipment. Registration is attained by assessment of the quality assurance system against the requirements of the appropriate DoH guide to good manufacturing practice (GMP) or, where no GMP exists, the relevant part of ISO9000. The GMPs are based on **BS5750**: 1987.

The procurement of supplies from registered manufacturers has been encouraged and already applied by most health authorities but, from January 1993, European directives on product certification (CE marks) and manufacturer registration have been effective. Reciprocal auditing arrangements have been operating for some years between the DoH and the Food and Drugs Agency covering the USA and Canada and also with SQS in Switzerland. Discussions are currently taking place on the European harmonization of standards but as yet no agreements have been reached.

On the wider front of quality improvement, the emphasis has been on better service and value for money. The DoH is committed to developing partnerships with suppliers, introducing just-in-time philosophies using electronic data interchange (EDI). EDI is a paperless transaction already used in some NHS locations. This relies heavily on ensuring products and services are right first time, placing further stress on approval and monitoring of suppliers. The NHS is now making significant and increasing use of EDI. To communicate the quality and value message, the DoH is running conferences on the theme 'Selling to the NHS – meet the buyer'.

The DoH is also introducing performance measures across the organization. Some of these have received widespread publicity; for example, throughput of hospital beds and hospital waiting lists. These measures are being determined to help plan the resources needed to run the NHS efficiently. There are two areas where performance measures are being reviewed. The first is the degree of customer (patient) satisfaction. This can be assessed by questionnaires completed by the patients, indicating if they were satisfied with the standard of care. The second area is research into the effectiveness of different types of treatment in the longer term; this can also be expressed as the quality of life of the patient.


## Total Quality in the UK motor industry

Motor manufacturers are at the other end of the spectrum: very much involved in the manufacture of a product. What follows is an outline of typical activities promoting the Total Quality approach within the industry.

The motor trade is a major sector that creates demands on many organizations for specific improvement activities; the automotive industry is the largest manufacturing industry in Britain. The sector is represented by the Society of Motor Manufacturers and Traders Limited (SMMT), an association of some 1,400 member companies involved with the manufacturing or marketing of vehicles and components. The SMMT provides members with assistance in areas such as legal and consumer affairs; technical, trade and media relations; and economics and trade statistics. It also produces a wide range of publications, many being available to non-members. The British Motor Show is one of the events organized by the SMMT.

The SMMT promotes the 'built in Britain' message, believing quality has to be at the heart of how every business is run. The drive to improved quality has been channelled through its quality and reliability panel, which is made up of representatives of vehicle and major component suppliers. The SMMT has been associated with many quality initiatives. Its own booklet on the subject has sold over 20,000 copies, guidance documents on product liability have been produced and a publication on **failure mode and effects analysis** has been launched.

The quality and reliability panel has objectives which are reflected in its mission statement. The aims are two-fold: first, to encourage the UK automotive industry to produce and market better quality and more reliable products; and, secondly, to react to the actions of external forces where they affect quality. The panel represents the whole industry – not just the vehicle manufacturers, but also the component suppliers – and maintains close links with other external quality groups.

One of the panel's key objectives is to achieve closer understanding and awareness of quality and reliability concepts in the industry and to promote the adoption of common supplier assessment procedures. As a direct result of this the SMMT is supporting the recognition of ISO9000/EN29000/**BS5750** (international, European and British identical standards) and recommends that, whether companies contract work out or buy in supplies, they should accept no standard which is below the level laid down in ISO9000. This is a major development in an industry which has been dominated by quality standards specified by the vehicle manufacturers. However, some vehicle manufacturers which have additional requirements will continue to assess suppliers to their own standards, although it is likely that suppliers with ISO9000 approval will gain growing recognition. An increasing number of companies within the automotive industry are now becoming independently assessed to ISO9000 and others are expected to follow.

# Part Four

# Profiles of Success Through Total Quality

# Introduction

Lynne Henderson visited a number of UK organizations, each well known for a Total Quality approach, in search of the key factors in their success. Here is her report.

'When I was initially asked to profile organizations in the UK who had "successfully implemented TQM", I must admit that I felt like Edmund Hillary standing at the bottom of Mount Everest for the first time. I felt ''fit'' enough to accept the challenge but I knew that I was going to need the right tools to do the job! The tools for this job included a format for a profile and a list of general questions and prompts for the interviews.

'Deciding the route to take was the easy part. I needed to climb as high as I could in each organization and I needed to cover as many different types of industrial terrain as possible. Not all paths were easy and sometimes I had to turn back and go another way because I could not meet certain organizations in my time-scales or, sometimes, I came across organizations who did not want to participate for varying reasons. The twenty visits that I did carry out brought with them such a high level of enthusiasm from the companies involved that it made the challenge so much more enjoyable. They looked upon the visit as an opportunity to ''share their success'' with other companies rather than ''letting out their quality secrets''.

'Sharing best practices or good communication is a part of TQM and a fundamental part of the cultures that these companies have invested time to develop. It was a breath of fresh air to experience organizations who emphasize that the core of the way they work is openness and honesty. TQM needs good leadership based on a culture that lends itself to new ideas, committed people at *all* levels in the company and methods of working that are flexible enough to undergo change.

'Of the twenty companies that I visited we have chosen eight for this book – not necessarily because they were seen to be the best but because they are a representative sample. They provide you with a snapshot of the effort that these companies have placed behind their quality initiatives. Some of the initiatives described have been operating for well over a decade, but Total Quality doesn't happen overnight. I hope that these profiles provide you with that extra encouragement to keep driving forward with your own projects.

'I reached the top of the mountain tired but exhilarated! These companies, by example, are shouting the way forward from their rooftops and yet there are still organizations whose own whining drowns out the message. I'm so glad that I climbed the mountain – *I can look back and still see the way forward with Total Quality.*'

In the profiles which follow, terms which are described in detail in the Executive Encyclopaedia appear in bold type. We are grateful to the companies concerned for allowing outlines of their successes in Total Quality Management to be included in this book.

# Boots The Chemists

*'Clarifying processes and procedures has been our most time-consuming – and most valuable – work.'*

## Company profile

Unique among UK multiple retailers, Boots The Chemists has maintained a substantial inhouse facility to provide and maintain the company's infrastructure. This store planning facility has a staff of 350, responsible for £135m of company expenditure. Its activities include creative design, planning, project management and implementation of a substantial capital development programme, store cleaning and a facilities management service for a property portfolio in excess of 2000 sites.

## Change implementation ... how it was done

Before TQM implementation, the Store Planning department faced a major problem – a reputation for being 'big, slow, expensive and inscrutable'.

The Store Planning group's senior management team saw TQM as a vehicle for change. Analysing their business in great depth, they built up, not a series of 'nice to have' changes that TQM could effect, but a business case for change which, when implemented, gave immediate benefits – a step change improvement.

The balance of the evidence came from the application of an essential technique – **Business Process Analysis** – which identifies and traces the key processes across the business. This analysis highlighted the need for simplification in store planning processes. Prime goals were improved customer focus, strengthened project management, the creation of performance measures and regular benchmarking – all to improve the bottom line.

Further contributions to the case for change were made by an analysis of the Cost of Quality – the cost of quality conformance weighed against the cost of non-conformance, such as abortive work, errors and rework. This amounted to a staggering £2.5m – or 25% of Store Planning's cost base – and 17% of this cost base was due to the cost of non-conformance.

Having defined these goals and identified the key value-adding processes needed to reach their targets, Store Planning has restructured its entire operation. A **Store Design Concept** for clients has been defined and standard 'models' applied in place of expensive, bespoke technical solutions. Project management processes have been established, a post-scheme 'after-care' system set in place and cost control re-evaluated.

### Summary of some of the major achievements attributed to the Total Quality initiative

- The department has relocated, and reduced its size by 13%.
- There are now fewer management layers within the department.
- Staff are empowered through Continuous Improvement Teams.
- The organization's services are now 'indispensable' and 'value-adding' according to its customers.
- The infrastructure is now in place to ensure continuous improvement.

# Cossor Electronics

*'To excel in the quality of product, service and manufacture by involving all employees, suppliers and customers in pursuit of company wide error free performance.'*

### Company profile

Cossor Electronics is an engineering company, designing and manufacturing specialist high technology electronic and electro-optic products for the defence and commercial markets. The company employs 1,150 staff at Harlow in Essex.

### Change implementation ... how it was done

In 1983 Cossor reviewed their quality management methods and found that they were lacking. A new approach to quality was required, one that would reduce pressure from customers and increase business efficiency. A strategy was put to the board of directors in 1983. Its approval saw the advent of the Total Quality Management initiative for Cossor Electronics. Cossor commenced formal quality improvement in 1984, with the objective of implementing a company-wide programme.

Cossor's route to quality improvement takes on board ten basic principles:

1  Management commitment – Cossor management believe that the success of a TQM programme starts with their demonstration of commitment. This commitment is continually repeated and reinforced to all the workforce.

2  Quality cost accounting – Cossor's improvement campaign began with a snapshot of **quality costs** which formed the basis for justification of the programme to the board of Cossor.

3  Senior quality improvement team – a top-level quality improvement team was initiated very early on. It is responsible for laying down the overall policy, targets and motivation. It has involved senior management from an early stage.

4  Quality measurements – test and inspection data are fully exploited at Cossor Electronics. Their centralized reporting systems keep management aware of potential problem areas. The quality measurement data encompass the collection of all defect data, local and central analysis, trend reporting, the identification of key problem areas, the production of data for problem-solving teams and monthly reporting.

5  Quality awareness – Cossor have found this one of the most difficult areas to implement because the company had previously experienced a 'blame' culture. They embarked on a major training campaign which involved

all Cossor staff. Dedicated posters, articles in magazines, trend charts in prominent positions and **teamwork** all helped to overcome the problem.

6 Training – extensive training was carried out in improvement techniques, quality costing, analysis methods, statistics, experimentation and problem solving. Training in all of these techniques started at the top and was cascaded throughout the company.

7 **Problem-solving** teams – Cossor set up three levels of problem-solving teams. First, 'corrective action teams' were responsible for the identification and solution of major problems. Secondly, 'task teams' were responsible for investigating sporadic problems and suggesting the procedures necessary for improvement. Finally, 'quality circles', consisting of volunteers, were responsible for identifying, analysing and recommending solutions to specific problems.

8 Improvement of facilities and procedures – this covers several areas such as the initiation of control to maintain the gains resulting from the quality programme, error-cause elimination, continual improvement of the quality management system, the use of technology for foolproofing and the provision of facilities which will resolve identified problems.

9 Recognition – Cossor's target is zero defects, and recognition or award schemes have been developed to reward staff for their achievements towards this goal.

10 Quality councils – established in individual functions within the company to provide a method of developing quality management practices in each area, for the examination of the associated quality performance and to allow for the setting of a quality improvement strategy for that area.

The first phase of the programme, which began early in 1984 and ended in later 1985, saved the company almost £1 million on quality-related costs. This first phase was based on the ten points above.

Subsequent activities have looked outside the business and involved customers and suppliers. For example, two particular customers were chosen to pilot Cossor's customer initiative. Objectives and targets were set, and corrective action teams involving the customer were set up to tackle particular problems. The results were better than predicted and the project was extended to other customers. Customer perception studies were extensively carried out through the complaints system, during installation and acceptance phases. Results were analysed and each customer was involved in an awareness programme on Cossor's quality improvement initiative. This has extended to working with major customers for the mutual improvement of each other's businesses.

To achieve customer satisfaction every time, Cossor knew that they needed to work closer with suppliers. The first move was to tighten their requirements on 'on-time delivery'. Letters were sent to all suppliers requesting a signed commitment to adhere to delivery time-scales. The target set by Cossor was not more than three days early and zero days late. The initial response from suppliers was positive. They regarded the 'quality window' as a challenge and recognized it as a positive step to improving their own competitive position.

### Summary of some of the major achievements attributed to the Total Quality initiative

- The successful operation of several types of problem-solving teams and approximately thirty-five quality circles.
- The running of eighteen quality awareness poster competitions and subsequent production and display of twenty-five posters – some of which are displayed in the Department of Trade and Industry's poster campaign literature.
- An ideas campaign which resulted in over 3,500 cost reduction or quality improvement suggestions.
- The training of more than 650 employees in problem solving, quality improvement and teamwork.
- Dramatic reduction of costs of poor performance.
- Positive improvements in customer satisfaction.

# Dowty

*'TQM eliminates waste and improves profitability. It also improves customer satisfaction both internally and externally, which gives competitive advantage, which in turn leads to more business.'*

## Company profile

Dowty is the aerospace 'arm' of TI Group plc. TQM was initially implemented in Dowty Fuel Systems, whose core products are mechanical fuel metering systems, pumps and actuating systems, as well as the gas turbine engines that power commercial and military aircraft.

## Change implementation ... how it was done

Prior to TQM, the company's management were complacent and the customers dissatisfied. The company failed to win significant business on the European Fighter because it did not listen to its customers. The Managing Director believes that, although a major blow at the time, it was the best thing that could have happened to the company in terms of changing attitudes.

A critical mass of Dowty Fuel Systems' senior executives 'bought in' to the TQM process from the outset, following an off-site awareness day for the company's entire senior team, which outlined the bottom line benefits of the process. From this group of convinced executives, a steering committee was set up, which defined quality policy and the values and culture necessary to achieve it. It was recognized that early successes were vital to ensure that all staff 'bought in' to the practical benefits of the TQM process.

Analysis of the **Cost Of Quality (COQ)** was one of the first TQM executive techniques utilized by Dowty Fuel Systems' steering committee. This is a measurement of the cost of conformance, which is the deliberate investment made to ensure good quality, weighed against the cost of non-conformance – the price of getting it wrong. A £50m business, the company discovered that £6m was being lost through non-conformance – and this was a conservative estimate.

COQ was broken down into several key areas – scrap and rectification, excess and obsolete inventory, indirect rework, non-productive engineering, receivables and others, and delivery performance. Just four weeks after analysing COQ, **Corrective Action Teams (CATs)** were put in place by the committee for each of these areas, determining key processes and methods for improvement, and implementing performance measures. This swift action was a vital part of the plan to gain quick results.

The CATS were each set a major objective. The machining CAT, for example, was required to halve errors in six months – a tough task for the most error-prone part of the business. Trained in problem-solving techniques and statistical process control, the team was able to meet its target in the required six months; after ten months they had achieved an 82% improvement. The achievements of the team proved to be a valuable marketing tool, both in terms of convincing TQ sceptics and of convincing end customers that the company was serious about changing.

## Summary of some of the major achievements attributed to the Total Quality initiative

- In three years, Dowty Fuel Systems has improved already healthy margins by 50%.
- Again, in three years, the company's return on capital has trebled.
- Customer service has improved significantly.

# Express Engineering

*'Our primary goal today, and in the future, is to serve our customers, by providing quality products, delivered on time, at a reasonable price. WE WILL ACHIEVE THAT GOAL.'*

## Company profile

Express Engineering is a small subcontract company which employs around fifty people and has an annual turnover in excess of £2 million. The company manufactures a range of precision components and special-purpose equipment used by the Ministry of Defence and by aerospace and consumer product industries.

## Change implementation . . . how it was done

In 1983 the need for quality improvement was becoming obvious to the management of Express Engineering. Driven into action by a consultant's review on the standard of their quality, they decided that, in committing themselves to quality improvement, they would commit themselves to complete change.

A key initial step in their quality improvement programme was a 'customer perception' campaign. They spoke with customers about materials, packaging and service to facilitate better products and better response to customer expectations.

Monthly quality committees, consisting of staff from all levels in the company, were introduced. At each meeting, a different chairman – selected at random – is required to coin a quality slogan for the month ahead. 'Some of the time we are not perfect, but all of the time we are trying to be!' is a favourite amongst staff, and was donated by the managing director when it was his turn to chair the meeting.

The company was assessed by the BSI to **BS5750**: Part 2/ISO9002 in 1987. At the beginning of the improvement initiative, there was an inherent resistance to the implementation of BS5750 by the staff. They felt that it would only introduce bureaucracy into an already adequate system. Management overcame the difficulty by meeting all the staff to tell them the reasons for certification and to discuss ways of achieving it as practically as possible.

Training of management and staff has been a major commitment over recent years, because 'high-tech' engineering to extremely close tolerances calls for highly skilled engineers. A large part of Express Engineering's success stems from in-house training, making full use of all available grants to train staff in necessary skills. The extensive programme includes training for every member of staff in techniques such as teambuilding, quality, business development, leadership and preparation for the Single European Market.

To identify and analyse areas for improvements, Express Engineering have implemented a **quality cost** system and bonus schemes for quality ideas from staff.

A supplier-based initiative has been introduced to ensure that those suppliers who were not BSI approved were made aware of the quality standards required by Express Engineering.

## Summary of some of the major achievements attributed to the Total Quality initiative

In 1987 Express Engineering achieved registration to BS5750: Part 2/ISO9002. The determination to achieve registration in a practical way was demonstrated by the institution of simple manual procedures and the imposition of the minimum of bureaucracy.

In 1988 the British Quality Award was presented to the company and HRH Prince Charles officially opened their new factory.

# Gwent Health Authority

*'Every person is responsible for the quality of service that they provide.'*

## Company profile

Gwent Health Authority supplies health care needs for the 500,000 population of Gwent. It employs 10,800 staff and is funded by the National Health Service.

## Change implementation . . . how it was done

In 1986 the chief administrative nursing officer was appointed director of quality assurance and was given the task of formulating a strategy for quality improvement. It was soon realized that in order to make a campaign work, it was necessary to learn from those who had already started down the Total Quality route. The year 1987 saw the appointment of the quality officer and health economist and the start of a staff awareness campaign.

The training and awareness built up over the year led to a mission statement of 'best care, cost effectively' being launched. Widening the 'ownership of quality' became the objective of the next stage of the training campaign. Staff training in aspects such as setting performance standards, quality groups and the benefits of participation in task forces began.

A quality circle campaign has been a successful district initiative. An additional bonus obtained through the work of one circle was its success in reuniting mentally handicapped people with relatives whom they had not seen for many years.

As a result of a consumer satisfaction survey, support groups have been set up which bring together experienced staff, doctors and families caring for the old and infirm so that they can share problems and solutions. The result is better care for the elderly people and for their carers.

An external feedback was seen to be needed. Complaints were closely monitored and objective corrective action set. Senior managers were heavily involved in this part of the project and targets were set to answer all complaints within twenty-four hours of their receipt and resolve them within three weeks. All complainers were visited at their home by a member of senior management who was responsible for ascertaining the exact problem and giving the person asssurance of commitment to resolve the problem.

A major investment in a computer system has allowed staff to monitor costs and complaints using **Pareto analysis** techniques.

Ongoing customer feedback has proved essential and an 80% response rate has been achieved with every customer survey. This has allowed staff to review and improve customer care and operational efficiencies.

A snapshot of **quality costs** was taken in one of the pathology laboratories as an experiment in the use of the technique. It was based on 'the cost of infection to 100 patients'. The criteria for the study were: the cost of care, cost of period in hospital, cost of drugs to cure the problem and the cost of not being at home with families. The first snapshot was extremely successful and provided problem-solving task teams with objective evidence of the need for a preventing approach.

A initiative is under way to look at suppliers and their methods of distribution throughout the district, so that drugs and equipment are brought to patients in a more timely and reliable way. Standards are being set, and as a result the quality of the service achieved by suppliers and procurers can be measured.

The district adheres to the philosophy that changes cannot be implemented without the support of staff at all levels. This is increasingly being achieved by the involvement of all staff in the setting of standards for quality. The quality of service achieved is then measured against the outcome objectives, which are always written in a way which allows audit.

## Summary of some of the major achievements attributed to the Total Quality initiative

Morale has improved. Staff within the district feel less isolated and believe that there now exists a team spirit within the health care organizations in Gwent. To help with the awareness and team spirit, an in-house magazine called 'Quality Grapevine' was launched. It has had the particular benefit of getting to people who are difficult to group into teams and ensuring that the message is continually reinforced.

# Hewlett-Packard

*'In order to compete in a global economy, our products, systems and services must be of a higher quality than our competition. Increasing Total Quality is our number one priority here at Hewlett-Packard.'*

John Young, President and Chief Executive Officer

## Company profile

Hewlett-Packard designs, manufactures, markets and services electronic business systems, devices and peripherals for a wide range of industries. In the UK the company employs over 3,500 staff and has a turnover in excess of £450 million per annum.

## Change implementation . . . how it was done

Hewlett-Packard's traditional strength has been to design products which outperform those from competitors or which perform a totally new function. In today's market, they have recognized that this is not sufficient for survival. Customers are demanding products and support that are strategic to their business success. To maintain custom and market share, Hewlett-Packard have achieved closer working ties with customers through Total Quality Management.

As early as 1957, management at Hewlett-Packard held a meeting to discuss and lay out what was central to the way the company should operate. These concepts comprise 'The H-P way'.

- Belief in people – freedom
- Respect and dignity – individual self-esteem
- Recognition – sense of achievement
- Security – performance and development of people
- Insurance – personal worry protection
- Share benefits and responsibility – help each other
- Management by objectives – rather than by directive
- Informality – open communication
- A chance to learn by making mistakes
- Training and education
- Performance and enthusiasm.

Employees at every level within the organization have the opportunity, through a structured development plan, to upgrade or realign their skills and capabilities to match market requirements and personal expectations. The emphasis on training and development is supported by continuing in-service education, encouraging personal contribution and rewarding achievement. For example, all staff trained in Total Quality culture (TQC) skills (in an intensive four-day training course) are regularly monitored after the course by internal consultants who encourage and assist them to implement the techniques in their own processes.

A key thrust of H-P's initiative has been to ensure that managers understand their role for the 1990s. Managers are trained, and retrained, at off-site process of management (POM) workshops where they are encouraged to develop a strategy under five headings:

1  Develop a statement of purpose and direction – set a shared mission.
2  Build a shared vision – involve the team in building a shared vision for the department.
3  Develop shared plans – involve the team in developing plans to reach the vision.
4  Lead the course of action – provide resources to implement plans.
5  Evaluate results and processes – check success, continually improve.

POM was developed within H-P from a study of the managers. The study showed that the most successful managers get extraordinary things done through their people. POM training imparts these proven techniques to new managers to provide them with the skills to become effective. Ongoing internal surveys are used to assess management performance. These are issued to both staff and managers. Results are graphically displayed and analysed to set higher performance targets and improvements in process techniques for management.

The general manager of each division demonstrates total commitment to quality and each manager instigates a system to review four crucial areas:

● customer focus
● planning
● process management
● total participation.

Suppliers are also critical, and H-P's supplier initiative has led to single sourcing of suppliers and assisting them with the implementation of just-in-time and zero defect techniques. They have utilized the **Deming** approach throughout their improvement initiative to encourage ownership through training and awareness. H-P's philosophy on Total Quality is about three things: customer focus, people participation and process improvement.

The monitoring of customer satisfaction levels is a major activity of the company every day. Customer satisfaction surveys (CSS) are achieving a 50% response. The surveys are a scientific assessment of customer satisfaction and the fundamental measure of customer 'delight' in product and service.

In H-P the customer is the most important person in the process and is always right.

### Summary of some of the major achievements attributed to the Total Quality initiative

In recent years Hewlett-Packard has grown by 20% each year. Their unfaltering reputation has taken them from being one of the top 200 companies in the world, to being in the top 40. Their business approach has brought with it customers who regularly state that Hewlett-Packard offer the best facilities and staff of any company.

Internally, Hewlett-Packard have always been staff oriented and they have recognized that their new TQC training programmes and TQC teams have improved commitment to the quality culture. The company works as a team, holding communication meetings once per month to reinforce the unifying spirit achieved over many years.

Their quality products have quantifiably improved, and a recent study has shown that products are ten times less likely to fail today than they were five years ago.

In 1978 Yokogawa Hewlett-Packard in Japan was one of the lowest quality, lowest productivity divisions in the corporation. In 1982 they were awarded the Deming Prize for Quality. They had become the highest quality, highest productivity division in the corporation. Hewlett-Packard say that this was nothing to do with being Japanese. It was simply due to the management of quality through TQC and 'The H-P way'.

# IBM

*'Our aim is to provide products and services to satisfy our customers in terms of delivery, quality and function at the right price.'*

## Company profile

IBM employs 1,600 people at its Havant plant on a site covering over one million square feet. The plant designs and manufactures information storage products, telecommunication and finance systems.

## Change implementation . . . how it was done

One of the first actions undertaken was to establish how IBM's customers perceived the performance of their products. IBM was initially shocked to find that customers were not totally satisfied. Further investigation showed that improvements were required in packaging, integration with products not produced in Havant and simplicity of installation, and attention to simple details such as labelling was needed. These findings triggered off various strategies within the plant.

A steering committee, including all the senior management team, was set up. This group developed a strategy that would, ultimately, be driven from below, but would initially be given its momentum from the top. To meet this strategy, the plant's goals had to be re-evaluated and clearly defined. These were then cascaded down through the organization until every employee's individual objectives reflected their contribution towards achieving these goals.

A strategy for the introduction of continuous flow manufacturing (just-in-time or JIT) techniques was introduced in 1984. The strategy started with an education programme for all employees. When some results were evident, the strategy was extended to IBM's suppliers. An extensive education programme was established to brief suppliers' senior management at supplier seminars. This was the start of the supplier certification programme which is based on **BS5750** and covers process qualification, project management and statistical techniques.

In 1985–6 techniques known as 'process quality management' and 'departmental purpose analysis' (DPA) were introduced. The aim of DPA is to assess the non-value-added activities conducted by a department with the intention of eliminating them. The analysis proved successful – 25 % of activities were identified as non-value added and subsequently eliminated.

In addition to these activities, the company has introduced small multi-skilled teams to improve the relationship between manufacturing and design in order to ensure both correct product and process design. Improvement teams have been initiated across the plant for **problem solving** and the steering committee is responsible for the introduction of an award scheme presented for the 'best' project. It is known as the Director's Award and is presented quarterly by the plant director at a private luncheon. A further recognition programme, 'company visits', caters for those who constantly participate in improvement projects but who have never been a part of the Director's Award. Exchange visits of thirty to thirty-five people with other companies who have an acknowledged quality reputation are arranged every four months.

The company is also certified to AQAP 1 and BS5750: Part 1. However, while IBM believe that these standards are necessary for their business, they realize that the true route to quality is via their people.

### Summary of some of the major achievements attributed to the Total Quality initiative

IBM's quality initiative was built against a background of changing business and customer demands. The emergence of competition from the Far East has had a major impact on their business. It forced IBM to evaluate its standards against 'best of breed' competitors and to recognize that customer expectations had changed.

During the nine to ten years that the Havant plant has been running a quality programme, there has been a high level of investment in education and senior management time. The results of this investment are seen in a rolling improvement programme. Manufacturing cycle time has been reduced by 80%. Inventory value has been reduced by 66%. Inventory turnover has increased threefold and revenue per person has been increased by eight times. The programme has a high level of employee participation – 90%. Last year IBM Havant has attributed in excess of £10 million savings to it.

# National Westminster Bank

*'The NatWest way is to bring quality to our customers, investors, people and the community.'*

## Company profile

The National Westminster Bank plc is the largest High Street bank in the UK, offering its services from 3,100 branches in 23 regions.

## Change implementation . . . how it was done

Deregulation within European markets has heralded a new era of customer dominance over any bank's corporate affairs. NatWest has made quality service its top priority in the quest to retain a dominant market position.

Their quality programme is split into four main areas:

1 Staff awareness – to launch the programme, all managers attended a series of workshops away from their offices. The rationale behind the drive for service quality was given and they had the opportunity to present their own ideas and input.

   Following this, the entire UK-based staff attended quality service events. Almost 60,000 staff took part, over a period of 6 months. Attendance was mixed across all job levels. For example, at one session the bank chairman sat alongside new recruits just out of school. The day was divided into four sections and dealt with awareness, behaviour, personal skills and teamwork. Quality service awareness is considered to be an ongoing method of training. The bank recruits almost 10,000 people every year and believes that quality service awareness should be an integral part of their induction training.

2 Quality service action teams – these are quality circles made up entirely of volunteers and not usually involving management. The average size of the circles is six and they meet on average every fortnight. Their task is to address local quality issues using a six-stage process moving from problem identification to a suggestion of action plans for the implementation of corrective action. To date, teams have identified and implemented over 7,000 improvements and solutions.

3 Market research – market research is seen as an investment in customers. It ensures that the bank retains a focus on the customers' expectations of the operation and keeps an external driving-force behind anything that is done.

Competitive **benchmarking** continues to evaluate how the customer rates NatWest against competition.

4   Service standards – utilizing the market research information, NatWest has developed areas of performance-standard setting. These are formalized targets for achievements within key service areas, ranging from dealing with customer complaints to response time on particular services. Standards and their associated targets ensure that the bank is delivering what the customer expects.

### Summary of some of the major achievements attributed to the Total Quality initiative

The success of the programme is linked to the introduction of the Quality Action Awards. All UK-based parent and subsidiary company staff are eligible for these. They are presented to members of staff who have contributed a practical initiative with significant results – not just financial.

Grassroots ownership of the quality service programme has been achieved via the establishment of service standards and through the efforts of the quality service action teams. This local ownership of quality has become a powerful force for change and forms an integral part of a Total Quality Management programme through which NatWest has secured the most important contributors to success: the enthusiasm and commitment to the programme throughout every level of staff.

# Nissan

*'Total Quality is a continuous improvement process building quality in at every stage of design and production.'*

## Company profile

Nissan is a major and growing automobile manufacturer based on a 733-acre site in Sunderland.

## Change implementation . . . how it was done

Nissan likes to refer to itself as a British company with a Japanese flavour. Japanese techniques of quality management have been adapted to suit the British culture.

In this high-production environment Nissan has encouraged all staff to own the quality of their output. Staff are trained in many techniques of quality before being certified to operate on the main product. Training is carried out in at least three stages against an extensive training budget.

The overall philosophy at Nissan Motor Manufacturing is one of Total Quality control. This means that the theme of quality runs through every aspect of the business at all times. Quality is not something that is left to quality controllers, but is the responsibility of every single person at the plant and everyone has a significant contribution to make.

At the plant there are four main features in the process of continual quality control. First is the philosophy running through the production process, which is that no defect will pass from one part of the process to the next. In a series of neighbour checks, whenever one component or subassembly is to be passed to the next station or 'neighbour' it is checked by the operator. By building up this network the integrity of the manufactured quality is continuously developed within the process.

The second feature consists of quality targets which are divided into supervisor sections so that each team of people in the production process has a defined target to aim at and monitor. The targets, which are regularly enhanced to aim at a constantly improving product, are important in that they make everyone aware of their own individual contribution to the overall philosophy.

Third, Nissan have developed a visible management system, comprising large display boards in every supervisor's section. On these boards quality data and performance are displayed. The graphs and charts are completed by the individual sections and thus the awareness of the need for quality is continuously enhanced.

Finally, there are help lamps. A wire running the whole length of the production line can be pulled at any time by an operator who has a 'quality concern'. This activates a siren and a flashing light and draws the attention of a supervisor who can investigate the area of concern and take the appropriate action. The use of the lamps is positively encouraged as an aid to highlighting any quality concerns within the plant.

Customer initiatives have taken the form of direct contact. Questions like 'What was receipt quality like?' and 'What is it now?' are asked. Nissan employees believe that the company must be pro-active in terms of quality and customer care.

An extensive audit and review system is Nissan's method of reviewing the progress of the programme. Audits are carried out by quality assurance and production representatives. Everything in the company is subject to audit:

- customer feedback
- systems
- products
- people.

Each of these is audited once a month and subsequently results are reported.

Supplier initiatives are a major part of Nissan's programme. They have actively worked with the majority of suppliers, increasing awareness, zero defect deliveries and supplier development team activities. Each of the suppliers has gone through an extensive quality training programme to improve the effectiveness of their businesses. Nissan believe that success breeds success and, through a jointly developed approach, have built mutual trust through pre-production quality assurance activities and high quality of parts supplied.

### Summary of some of the major achievements attributed to the Total Quality initiative

- Productivity and plant capacity have increased.

- Higher morale and enthusiasm, and increased teamwork.

- Continuous improvement is everyone's responsibility. It is important to note that no direct financial reward for achievement is given and no suggestion schemes exist.

# Pitney Bowes

*'You have to create measures within all aspects of your business, ultimately leading to customer satisfaction. These will drive your business.'*

## Company profile

Pitney Bowes Inc, USA, is a $3 billion business, its prime market being postage meters and paper handling machines. Listed in the Fortune 500, the company has continued 10% growth. It was against this background that expectations for Pitney Bowes UK were set.

## Change implementation ... how it was done

Prior to TQM implementation, the UK company had a factory turnover of £11m, declining market share and lack of performance indicators. The parent company believed that the UK company's manufacturing standards were poor, and needed significant improvements in order to compete in a global market and to be a worldwide supplier of product. TQM was seen as a practical vehicle for change.

Having made the decision to implement TQM within their manufacturing division, Pitney Bowes' senior management team analysed their business and pinpointed two main areas which demanded investment – technology and people. Indeed, improvements through effective investment in people became the company's ultimate vision. Having created the vision, this had to encompass a set of strategic goals – increased market share, becoming a least cost provider, reducing time to market and, above all, a focus on customer satisfaction.

Using the key executive technique of **competitive benchmarking**, the company's senior executives compared their procedures with those of their international competitors, so identifying the areas in which they had to succeed. Further benchmarking, initially against their US mother company and secondly against best practice companies worldwide, enabled the manufacturing division to create standards and key measures. Using this information, they also set real targets, in inventory, product quality and lead time to market.

**Corrective Action Teams (CATs)** were introduced to address the company's key target areas. They identified processes and problems and introduced new processes, with appropriate measures in all areas. These highly visible measures, based on the teams' own solutions, encouraged ownership of the processes.

The company firmly believes that it is the ownership of such measures which matters, rather than the measures themselves. Prior to TQM implementation, the existing measures were perceived as management 'sticks', rather than 'carrots', and as such were useless. They were not 'owned' by the people who were doing the job. Now, there are measures on every area of Pitney Bowes' shop floor; staff are committed to their drive towards zero defects, and proud of their achievements.

The original vision of Pitney Bowes has been achieved. Staff are becoming results oriented and are demanding a different type of leadership, which provides training and support, but which grants them ownership of their processes.

### Summary of some of the major achievements attributed to the Total Quality initiative

- The company is now a £100m business.
- Work in progress is 100% lower, and throughput three times higher.
- Lead times have been reduced from six months to one month over the space of two years, with a two day lead time for special orders.
- On one of the production processes, stock turns have increased from two to eleven in the space of three months.
- There have been space savings of 60% and labour reduction of 30% – with the workforce themselves actually requesting labour reductions on the production line.

# Sony

*'Sony (UK) Limited, Bridgend, operate a documented, understood and maintained strategy of Total Quality control consistent with a policy objective of zero defects based on prevention rather than inspection, and the use of upstream action in problem solving.'*

## Company profile

Sony (UK) Limited manufactures colour television (CTV) sets for consumer and business markets worldwide. The company employs over 1,500 staff and achieves turnover in excess of £200 million per annum.

## Change implementation ... how it was done

In 1984 the company recognized that although they were producing good quality products, they were too inspection orientated and therefore inefficient with their resources. A business review was undertaken which identified three main reasons for changing:

- plant and products were becoming much too complex for current systems
- market competition was increasing dramatically
- production volumes were increasing from 500,000 sets per year to 1,000,000.

These factors demanded a fundamental reappraisal of the plant's business strategy and were integrated into a Total Quality approach.

The quality improvement programme was launched in 1984, with the following fundamental improvement targets:

1 The role and capability of management – each manager is encouraged to adopt the role of managing director for his/her own immediate work area. This concept was introduced top-down, starting with the production department and extending to other areas (down to supervisors within six months). Each key person was responsible for setting targets with his/her superior which were then reviewed weekly to evaluate performance. Where targets were met, new tougher ones were set. Where they were not met, lists of problems were identified and solutions found. 'Problem ownership' was a main theme; managers were encouraged to set up small participative **problem-solving** teams within their departments and, by the end of 1985, the company had evolved a successful quality circle programme.

2 The total involvement of its workforce – training of the entire workforce on problen solving-techniques – and vital use of communications systems improved quality awareness throughout the plant.

In particular, all were trained in a problem-solving technique called CEDAC (**cause and effect diagram** with the addition of cards). It is based on the Ishikawa method, but allows any personnel to contribute toward a solution. CEDAC diagrams are displayed in every work area and are thus visible to everyone for input. A management concept called 'top three meetings' was introduced. They were held at the end of every day to identify the top three problems of that day. The members of the meeting did not leave until they had found solutions to all three.

3  The design of the product – the objective of designing new models for maximum quality – took over as a major theme alongside features and cost. Production management was involved with design at the earliest stages to achieve this target. Advanced training was given to all personnel before any new product was released into production.

4  The company's relationship with its suppliers – 1987 saw a major attack on the quality of all incoming components. Work groups developed strong working relationships with external suppliers, moving away from incoming inspector to direct quality control at source. The incoming inspection department now agrees quality objectives with each supplier and regularly audits their processes to ensure defect-free work.

### Summary of some of the major achievements attributed to the Total Quality initiative

The improvement programme has increased motivation, determination and enthusiasm. **Quality costs** have been reduced by a factor of 3:1 and measurable mistakes per person have reduced from 500 to 20. The quality circle programme has been a major success and has seen the introduction of a recognition scheme for participants.

# Conclusion – The key factors of success

Each of the profiles shows an approach to Total Quality which has been successful in one particular set of circumstances. These organizations have unlocked the potential in their situation, their technology and, especially, their people through a Total Quality approach. Their choice of tools and techniques, adapted into their situation and developed with their own ideas, has released the benefits.

The fact that the approaches are different emphasizes that Total Quality Management is *not* a specific programme to be applied to all organizations in a standardized way. However, there are common threads – 'concepts', as described in Part Two of this book – that can be identified.

### Customers – internal and external

Each of the companies that was profiled understood that commitment to customers was fundamental to every business activity.

IBM says that it is no longer enough just to 'conform' to customer requirements – you must satisfy them in every aspect. Your customer's requirements and expectations may be two completely different things; successful companies recognize this and take steps to satisfy both.

### Never-ending improvement

Total Quality is not something that you can buy, or something that you can tick off as complete on your business plan. TQM is a commitment for life; it is the way that you run your business.

Hewlett-Packard, and others, call it 'continuous improvement'. They say that it is about continually improving people and processes – but doing it in such a way that you hold the gain and do not slip back into old habits.

### Control of business processes

Knowing your key processes and keeping control of your key processes means involving the people who know – those who do the job.

A Total Quality approach avoids writing a procedure which is subsequently handed out and then ignored by the people in the process because it does not reflect what is actually done! Before processes can ever be improved or controlled, they must be understood. Dr Deming calls that 'profound knowledge'.

Sony is a company which places the ownership of the process in the person who does it. This has led to excellent improvements being made in its larger business processes.

## 'Upstream' preventive management

We all understand that 'fixing things right' costs money and is very demoralizing. In TQM, we talk more about building quality into the service or product and so Total Quality means preventing problems in the first place.

NatWest gave a good example of prevention management. Banks are something that we all use on occasion. For those of us in business, that usually means we call there between 12.00 and 14.00 – during our lunchtimes.

NatWest avoids the problem of having customers standing in very long queues for long periods in their lunchtimes staring at the only two teller positions that are open. They make sure that between the hours of 11.30 and 14.30, *all* teller stations are open. This initiative did not come as a directive from the top, but as a suggestion from a quality circle group in one branch.

## Ongoing preventive action

Prevention management is something that we all need to work at because it often involves not only culture change but commitment from the company to provide the resource needed to maintain that preventive approach.

An excellent example of this can be seen at Express Engineering. There, the management team reinvest a substantial amount of profit in their company to make sure that their staff have the most up-to-date machines and tools to do the job well and have the best training to maintain their standards of excellence.

## Leadership and teamwork

Management by directive and management by suspicion are as far from TQM as chalk is from cheese. Cossor Electronics, Hewlett-Packard, IBM and, in fact, every company visited said that no manager can manage from behind a door or a desk. Managers need to be leaders. They must coach, develop, remove barriers, create an environment of innovation and creativity. They need to motivate their teams and, above all, work with them, not against them. Autocracy is one poison not tolerated in TQM.

Hewlett-Packard have had a culture of teamwork since the company's inception fifty years ago. They practice MBWA – management by wandering around – which means knowing their people, their capabilities, their strengths and leading the team toward their shared goal.

These experiences with the basic concepts can show the way to your Total Quality approach. Part Two of this book guides you in interpreting these concepts, developing a management framework and planning the launch of a change to Total Quality Management – you may want to go back to it now you have studied the examples in this part. We wish you well.

# About the major contributors

This Peratec executive briefing relays the benefits, pitfalls and successes of implementing Total Quality processes that Peratec's clients have experienced. It contains detailed contributions from the following individuals and organizations.

| Contributing Organizations | Key Authors |
| --- | --- |
| Hewlett-Packard | Wally Cichorz |
| National Westminster Bank | Ian Ferguson |
| Nissan | Hugh Gallacher |
| IBM | Chris J. Hakes |
| Sony | Robert McLellan |
| Gwent Health Authority | David Spilsbury |
| Express Engineering | David Walker |
| Cossor Electronics | Lynne Watson |
| American Association for | |
|    Quality and Participation | |
| American Supplier Institute | |
| Boots The Chemists | |
| Dowty | |
| Pitney Bowes | |

## About Peratec

Peratec is the consulting arm of Pera International, carrying out all aspects of consulting and training activities.

Committed to improving its clients' business performance through an innovative pragmatic approach, Peratec's consultants deliver measurable results and a service tailored to the individual needs of each client.

Focusing its attention on people, business and technology issues, the company has in recent years enjoyed significant growth in demand for its consulting services in the areas of Total Quality Management, Quality Systems, Change Management, Human Resources Development, Applied Materials Technology, Strategic Marketing, Mergers & Acquisitions, Product Design, Technology Management, Design & Development and Manufacturing Management.

Peratec's Total Quality Management team is one of the largest and most experienced in the UK. It has an impressive record of helping companies to implement specific quality improvements or to take a more strategic approach to quality.

If you would like to know more about how Peratec can help you to improve your organization's performance in the market-place, contact:

| Peratec Limited | Peratec Limited | Peratec Limited |
|---|---|---|
| Pembroke House | Trident House | Melton Mowbray |
| Lydiard Millicent | 175 Renfrew Road | Leicestershire |
| Swindon | Paisley | LE13 0PB |
| Wiltshire | Strathclyde | |
| SN5 9LS | PA3 4EF | |
| | | |
| Tel.: 0793 772555 | Tel.: 041-8897876 | Tel.: 0664 481101 |
| Fax: 0793 770183 | Fax: 041-8878105 | Fax: 0664 501551 |

# Bibliography

Caplan, F. (1980) *The Quality System: A sourcebook for managers and engineers*. Chilton, Radnor, PA.

Conti, T. (1993) *Building Total Quality*. Chapman & Hall, London.

Crosby, P.B. (1979) *Quality is Free*. McGraw-Hill, New York.

Crosby, P.B. (1984) *Quality without Tears*. McGraw-Hill, New York.

Crosby, P.B. (1986) *Running Things*. McGraw-Hill, New York.

Cullen, J. and Hollingum, J. (1987) *Implementing Total Quality*. IFS (Publications) Ltd, Bedford.

Deming, W.E. (1986) *Out of the Crisis*. MIT Center for Advanced Engineering Studies, Cambridge, MA.

Feigenbaum, A.V. (1983) *Total Quality Control, 3rd edn*, McGraw-Hill, New York.

Garvin, D.A. (1988) *Managing Quality: The strategic and competitive edge*. The Free Press, New York.

Goldsmith, W. and Clutterbuck, D. (1985) *The Winning Streak*. Random House, New York.

Guttman, I., Wilks, S.S. and Hunter, J.S. (1971) *Introductory Engineering Statistics, 2nd edn*, John Wiley, New York.

Harrington, H.J. (1987) *The Improvement Process*. Quality Press, Milwaukee, WI.

Imai, M. (1986) *Kaizen: The key to Japan's competitive success*. Random House, New York.

Ishikawa, K. (1982) *Guide to Quality Control, 2nd edn*, Asian Productivity Organization, Tokyo.

Ishikawa, K. (1985) *What is Total Quality Control*. Prentice-Hall, Englewood Cliffs, NJ.

Juran, J.M. (1962) *Quality Control Handbook*. McGraw-Hill, New York.

Kume, H. (1985) *Statistical Methods for Quality Improvement*. Chapman & Hall, London.

Lochner, R.H. and Matar, J.E. (1990) *Designing for Quality*. Chapman & Hall, London.

Murdoch, S. (1979) *Control Charts*. MacMillan, London.

Oakland, J. (1986) *Statistical Process Control*. Heinemann, Oxford.

Ott, L. (1988) *An Introduction to Statistical Methods and Data Analysis*. PWS Kent, Boston.

Peters, T.J. and Austin, N.K. (1985) *A Passion for Excellence*. Random House, New York.

Peters, T.J. and Waterman, R.H. (1982) *In Search of Excellence*. Harper and Row, New York.

Popplewell, B. and Wildsmith, A. (1989) *Becoming the Best*. Gower, London.

Soin, Sarv Singh (1992) *Total Quality Control Essentials*. McGraw-Hill, New York.

Spenley, P. (1993) *World Class Performance Through Total Quality*. Chapman & Hall, London.

Taguchi, G. (1986) *Introduction to Quality Engineering*. Asian Productivity Organization, Tokyo.

Taguchi, G. (1987) *System of Experimental Design* (Vols 1 and 2). Quality Resources, White Plains, NY.

Taguchi, G., Elsayed, E. and Hsiang, T.C. (1988) *Quality Engineering in Production Systems*. McGraw-Hill, New York.

Walton, M. (1986) *The Deming Management Method*. Quality Press, Milwaukee, WI.

# Other sources of help

In addition to the consultancy, training and general support provided by our own organization, further guidance is available from the following bodies.

### British Standards Institution (BSI)

The BSI is responsible for preparing British Standards, which are used in all industries and technologies. It also represents British interests at international standards discussions, ensuring that European and worldwide standards will be acceptable to British industry. Some 70% of BSI standards work is now geared to international alignment. BSI also takes a prominent part in developing internationally agreed criteria for quality assessment and certification.

British Standards are compiled by committees representing those organizations with a close interest in the subject: manufacturers, users, government bodies, and safety, research and consumer organizations. There are now over 10,000 publications listed in the *BSI Catalogue*, including the key standard for quality systems, **BS5750**. Information about all aspects of BSI's work, including sales information on British, international and foreign national standards, can be obtained from:

Enquiry Section
BSI
Linford Wood
Milton Keynes MK14 6LE
Tel.:   0908 221166
Telex: 825777

General information and publications about standards, including teaching packs and visual aids for use in technical and consumer education, are available from:

Public Relations Department
BSI
2 Park Street
London W1A 2BS
Tel.:   071-629 9000
Telex: 266933

## *National Accreditation Council for Certification Bodies (NACCB)*

The NACCB was launched in June 1985 to make recommendations to the Secretary of State for Trade and Industry concerning the competence and impartiality of certification bodies, leading to their accreditation by the Secretary of State. Accreditation enables certification bodies to display the National Accreditation Mark alongside their own certification marks.
    Enquiries may be addressed to:

Secretary (NACCB)
Second Floor
3 Birdcage Walk
London SW1H 9JH
Tel.:   071-222 5374

## *National Measurement Accreditation Service (NAMAS)*

The NAMAS was formed in September 1985 by the amalgamation of the British Calibration Service (BCS) and the National Testing Laboratory Accreditation Scheme (NATLAS). It is responsible for the assessment and accreditation of laboratories carrying out calibrations, measurements or tests in support of product certification, procurements, regulatory bodies and quality assurance generally.
    Enquiries may be addressed to:

NAMAS Executive
National Physical Laboratory
Teddington
Middlesex TW11 0LW
Tel.:   081-977 32222 ext. 3672

*Association of Certification Bodies (ACB)*

The ACB was formed in 1984 with the purpose of working towards the improvement of product quality in British industry by representing the interests of its members, and co-operating with British industry and the government in achieving this objective. Initially the association has provided a forum where certification bodies can develop a collective input to issues associated with national accreditation of certification bodies.

Enquiries may be addressed to:

Mr R. Brockway (Secretary ACB)
BSI
2 Park Street
London W1A 2BS
Tel.: 071-629 9000

*The Institute of Quality Assurance (IQA)*

The IQA is a professional body established to serve the needs of all those involved with quality assurance.

It has been recognized for many years that product and service quality cannot be achieved solely by professional quality managers or technologists. It is essential that all managers should ensure that the quality function of their own departments is effectively fulfilled. Senior executives, in particular, must be committed to developing and implementing an overall company quality policy.

To further this end, the institute organizes short courses, seminars and conferences on all aspects of quality, covering a variety of industries and disciplines. It administers a structured training course programme and has details of the colleges which offer part-time and full-time courses based on the Institute's examinations syllabus. Details of these and of the postgraduate course on quality assurance and management are available from the Institute.

The Institute also publishes journals regularly, and has a comprehensive book list of publications devoted to the many aspects of quality assurance.

In all its activities, the Institute seeks to provide information suitable for all management and supervisory personnel.

General enquiries may be addressed to:

The Institute of Quality Assurance
PO Box 712
61 Southwark Street
London SE1 1SB
Tel.: 071-401 7227

## *The British Quality Foundation (BQF)*

The BQF, formerly the British Quality Association (BQA), exists to promote a better understanding of quality throughout industry and commerce. The establishment of industrial sector quality groups is an important part of its work. A number of these groups have already been set up, and are actively developing methods and techniques designed to improve quality in their field.

Membership is open to all industrial, commercial and other corporate organizations and is by no means confined to quality professionals. The aim is to make all businessmen and managers aware that the quality of their products or services is a vital factor in ensuring the competitiveness and profitability of their company.

The address is as for the IQA above; tel.: 071-401 2844.

## *The National Quality Information Centre (NQIC)*

The NQIC has been set up to help industry and commerce obtain the information they need to improve quality in their corporate activities and in the products and services they provide.

The NQIC is run by the IQA with support from the Department of Trade and Industry, and draws on the wealth of knowledge, and experience of quality management that exists within the institute and the BQF. It also has access to information held by many other organizations both in this country and overseas.

The address is as for the IQA above; tel.: 071-401 7227.

## *Association of Quality Management Consultants (AQMC)*

The AQMC is a self-regulating body whose members, by their training, qualifications and experience, meet requirements designed to establish that they are professionally competent to act as consultants to industry and commerce on a whole range of quality management activities.

Enquiries may be addressed to:

Honorary Secretary
4 Beyne Road
Olivers Battery
Winchester
Hampshire SO22 4JW
Tel.: 0962 64394

*The European Foundation for Quality Management (EFQM)*

The EFQM exists to support those European organizations wishing to create conditions to enhance the position of European industry (products and services) in the world market, by strengthening the role of management in quality strategies.

The EFQM is developing specific awareness, management education and motivational activities in close co-operation with other European organizations, aimed at management of companies rooted in Western Europe who are in search of quality management.

Affiliation with the EFQM is open to companies based in Western Europe with a declared top management commitment to its mission and objective. Under the same conditions of commitment, affiliation is also open to supporting organizations and higher institutes of learning.

Enquiries may be addressed to:

European Foundation for Quality Management
Building 'Reaal' Fellenoord 47a
5612 AA Eindhoven, The Netherlands
Tel.: + 31 40 461075
Fax: + 31 40 432005

# Index